High School
Algebra II
UNLOCKED

Penguin
Random
House

By Theresa Duhon and the Staff of The Princeton Review

The Princeton Review
24 Prime Parkway, Suite 201
Natick, MA 01760
E-mail: editorialsupport@review.com

Published in the United States by Random House LLC,
New York, and in Canada by Random House of Canada, a
division of Penguin Random House Ltd., Toronto.

Terms of Service: The Princeton Review Online
Companion Tools ("Student Tools") for the retail books
are available for only the two most recent editions
of that book. Student Tools may be activated only
twice per eligible book purchased for two consecutive
12-month periods, for a total of 24 months of access.
Activation of Student Tools more than twice per book
is in direct violation of these Terms of Service and may
result in discontinuation of access to Student Tools
Services.

ISBN: 978-1-101-92007-7
eBook ISBN: 978-1-101-92008-4
ISSN: 2471-3031

ACT is a registered trademark of ACT, Inc. and SAT is a
registered trademark of the College Board, neither of
which is affiliated with The Princeton Review.

The Princeton Review is not affiliated with Princeton
University.

Editor: Aaron Riccio
Production Artist: Deborah A. Silvestrini
Production Editors: Liz Rutzel and Harmony Quiroz
Template Design: Maurice Kessler

Printed in the United States of America on partially
recycled paper.

10 9 8 7 6 5 4 3 2 1

1st Edition

EDITORIAL
Rob Franek, Senior VP, Publisher
Casey Cornelius, VP Content Development
Mary Beth Garrick, Director of Production
Selena Coppock, Managing Editor
Meave Shelton, Senior Editor
Colleen Day, Editor
Sarah Litt, Editor
Aaron Riccio, Editor
Orion McBean, Editorial Assistant

RANDOM HOUSE PUBLISHING TEAM
Tom Russell, Publisher
Alison Stoltzfus, Publishing Manager
Jake Eldred, Associate Managing Editor
Ellen Reed, Production Manager

Acknowledgments

I first and foremost thank my family for always being supportive of me in all my endeavors. My high school math teachers, especially the talented Ed Davis, were inspiring and encouraging, as was Ali, who believed I was capable of studying graduate level math as an undergrad. I am grateful to Sara for opening the door to the world of test preparation work for me. Thanks to Aaron Riccio and the rest of the staff at The Princeton Review for the tremendous amount of work they put into the creation of this especially challenging book. I appreciate all of the opportunities TPR has given me over the years. Finally, I thank my wonderful husband, Takashi, whose support and confidence have sustained me throughout this process.

—Theresa Duhon

Additionally, The Princeton Review would like to emphatically thank Theresa Duhon for her unwavering, above-and-beyond commitment to this project, Maurice Kessler for his outstanding attention to detail and design, Craig Patches and Scott Harris for bearing with us through the art creation process, and Deborah A. Silvestrini for her nonstop, can-do attitude toward laying out this book. Special thanks also to Liz Rutzel and Harmony Quiroz for reviewing this text.

Contents

Register Your Book Online! viii

About This Book. x

1 Complex Numbers and Polynomials 1

Lesson 1.1 Complex Numbers. 2

Lesson 1.2 Operations with Polynomials 9

Lesson 1.3 Polynomial Identities. 13

Lesson 1.4 Graphing Polynomials (Beyond Quadratics). 17

Lesson 1.5 Tips for Factoring Polynomials. 24

Lesson 1.6 Polynomials in the Real World 31

2 Systems of Equations and Rational Expressions 41

Lesson 2.1 The Fundamental Theorem of Algebra 42

Lesson 2.2 Systems of Equations. 46

Lesson 2.3 Variation Equations. 49

Lesson 2.4 Rational Expressions. 54

3 Radical and Rational Equations and Inequalities 67

Lesson 3.1 Rational Equations in One Variable. 68

Lesson 3.2 Rational Inequalities in One Variable. 75

Lesson 3.3 Rational Functions. 78

Lesson 3.4 Radical Equations in One Variable. 92

Lesson 3.5 Radical Inequalities in One Variable 95

Lesson 3.6 Radical Functions 98

4 Trigonometric Functions 117

Lesson 4.1 Trigonometric Ratios and Triangles. 118

Lesson 4.2 Radians. 122

Lesson 4.3 Trigonometric Functions and Circles 128

Lesson 4.4 Graphing Trigonometric Functions 135

Lesson 4.5 Trigonometric Function Identities 150

Lesson 4.6 Trigonometric Functions in Real Life 158

5 Logarithms 175

Lesson 5.1 Inverse Functions. 176

Lesson 5.2 Logarithms 182

Lesson 5.3 Logarithmic Identities. 192

Lesson 5.4 Logarithmic Functions 198

Lesson 5.5 Logarithms in the Real World 208

6 More Functions 225

Lesson 6.1 Cube Root Functions 226

Lesson 6.2 Piecewise-Defined Functions 235

Lesson 6.3 Exponential Functions 252

7 Making and Using Mathematical Models. 273

Lesson 7.1 Manipulating Equations 274

Lesson 7.2 Interpreting Equations, Tables, and Graphs 277

Lesson 7.3 Modeling Situations with Equations and
Inequalities 285

Lesson 7.4 Modeling Situations with Graphs 289

8 Inferences and Conclusions from Data 303

Lesson 8.1 Data Collection. 304

Lesson 8.2 Means and Measures of Variability. 312

Lesson 8.3 Frequency Distributions. 316

Lesson 8.4 Probability Distributions. 325

Lesson 8.5 Sample Proportions and Sampling
Distributions 341

Lesson 8.6 Confidence Intervals and Margins of Error . 347

Register Your

1 Go to **PrincetonReview.com/cracking**

2 You'll see a welcome page where you can register your book using the following ISBN: 9781101920077.

3 After placing this free order, you'll either be asked to log in or to answer a few simple questions in order to set up a new Princeton Review account.

4 Finally, click on the "Student Tools" tab located at the top of the screen. It may take an hour or two for your registration to go through, but after that, you're good to go.

If you are experiencing book problems (potential content errors), please contact EditorialSupport@review.com with the full title of the book, its ISBN number (located above), and the page number of the error. Experiencing technical issues? Please e-mail TPRStudentTech@review.com with the following information:

- your full name
- e-mail address used to register the book
- full book title and ISBN
- your computer OS (Mac or PC) and Internet browser (Firefox, Safari, Chrome, etc.)
- description of technical issue

Book Online!

Once you've registered, you can...

- Access and download "Key Points" review sheets for each chapter.

- Work through additional Locksmith sample problems.

- Check for any corrections or updates to this edition.

Look For These Icons Throughout The Book

 Online Supplements

 Online Practice

The
Princeton
Review®

About This Book

WHY HIGH SCHOOL UNLOCKED?

It might not always seem that way, especially after a night of endless homework assignments, but high school can fly by. Classes are generally a little larger, subjects are more complex, and not every student has had the same background for each subject. Teachers don't always have the time to re-explain a topic, and worse, sometimes students don't realize that there's a subject they don't fully understand. This feeling of frustration is a bit like getting to your locker and realizing that you've forgotten a part of the combination to open it, only there's no math superintendent you can call to clip the lock open.

That's why we at The Princeton Review, the leaders in test prep, have built the *High School Unlocked* series. We can't guarantee that you won't forget something along the way—nobody can—but we can set the tools for unlocking problems at your fingertips. That's because this book not only covers all the basics of *Algebra II*, but it also focuses on alternative approaches and emphasizes how all of these techniques connect with one another.

How to Use This Book

The speed at which you go through this material depends on your personal needs. If you're using this book to supplement your daily high-school classes, we recommend that you stay at the pace of your class, and make a point out of solving problems in both this book and your homework in as many ways as you can. This is the most direct way to identify effective (and practical) tools.

If, on the other hand, you're using this book to review topics, then you should begin by carefully reviewing the Goals listed at the start of each chapter, and taking note of anything that seems unfamiliar or difficult. Try answering some of the example problems on your own, as you might just be a little rusty. Applying math skills is a lot like riding a bike, in that it comes back quickly—but that's only true if you learned how in the first place. Take as much time as you need, then, to connect with this material. As a real test of your understanding, try teaching one of these troublesome topics to someone else.

Ultimately, there's no "wrong" way to use this book. You wouldn't have picked this up if you weren't genuinely interested, so the real key is that you remain patient and give yourself as much time as you need before moving on. To aid in this, we've carefully designed each chapter to break down each concept in a series of consistent and helpful ways.

Goals and Reflect

Each chapter begins with a clear and specific list of objectives that you should feel comfortable with by the end of the chapter. This allows you not only to assess which sections of the book you need to focus on, but also to clarify the underlying skills that each example is helping to demonstrate. Think of this sort of goal-based structure as a scavenger hunt: it's generally more efficient to find something if you know what you're looking for.

Along those lines, each chapter ends with an opportunity to self-assess. There's no teacher to satisfy here, no grade to be earned. That said, you're only hurting yourself if you skip past something. Would you really want to jump into the pool before learning how to swim? You've got to determine whether you feel comfortable enough with the foundational understanding of one chapter before diving deeper.

Review

Review boxes serve as quick reminders of previously learned content—formulas and definitions—that are now being expanded on, or utilized to demonstrate new techniques. These appear at the start of each lesson, and are a bit like the lists of ingredients provided for a recipe, in that they ensure that you have everything you need to make (learn) something new.

Examples

Each lesson is hands-on, filled with a wide variety of examples on how to approach each type of problem. You're encouraged to step in and solve the problem yourself at any time, and each step is clearly explained so that you can either compare techniques or establish a successful strategy for your own coursework. Moreover, examples gradually increase in difficulty throughout the chapter. In moving from basic expressions to more complicated real-world examples, we hope to provide a clearer picture of how to break down concepts to their core techniques.

Locksmith Questions

This element features sample questions in the style of the ACT and SAT, providing you with an enhanced opportunity to see how these topics might appear on a high-stakes test. Answers and explanations appear separately on the next set of pages, allowing you to see how the skills and strategies that you're developing can be adapted or focused, regardless of the context in which a question appears.

Key Chains

Our Key Chain sidebars operate much like your actual key chains—they help to sort ideas and, through repetition and connection, to help ensure that you don't misplace them. After all, it doesn't matter how many keys you own if you can't find them when you need them! If you have the time, Key Chains encourage you to go back to earlier questions and lessons with new insights; if not, they double as mini-Review boxes, helping to remind you of all the applicable skills at your disposal.

End-of-Chapter Practice Questions and Solutions

It can be tricky to accurately assess how well you know a subject, especially when it comes to retaining information. We recommend that you wait a day or two between completing a chapter and tackling its corresponding drill so that have a good measure of how well you've absorbed its contents. These questions intentionally scale in difficulty, and in conjunction with the explanations, help you to pinpoint any remaining gaps.

Key Points

For additional support, we've placed printable key points online. These handy tables summarize the major formulas and concepts taught in each chapter, and serve as excellent review material.

WHAT NEXT?

Many of these Unlocked techniques can be applied to other subjects. If you're planning on taking the SAT or ACT, we recommend picking up a book of practice questions or tests so that you can keep those keys nice and sharp. If you're moving on to other courses, or higher-level AP classes, remember the connective strategies that most helped you in this book. Learning how to learn is an invaluable skill, and it's up to you to keep applying that knowledge.

Chapter 1
Complex Numbers and Polynomials

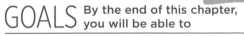

GOALS By the end of this chapter, you will be able to

- Perform arithmetic operations with complex numbers

- Perform arithmetic operations with polynomials

- Use polynomial identities to solve for certain quantities, find Pythagorean triples, and factor polynomials

- Graph polynomial functions and derive polynomial equations from given graphs

- Factor higher-degree polynomials completely, into any real- and/or imaginary-number roots

- Create and interpret polynomial functions and their graphs to represent and explain real-world relationships

Lesson 1.1
Complex Numbers

REVIEW

COMMUTATIVE PROPERTY OF ADDITION

$a + b = b + a$

ASSOCIATIVE PROPERTY OF ADDITION

$(a + b) + c = a + (b + c)$

DISTRIBUTIVE PROPERTY

$a \cdot (b + c) = a \cdot b + a \cdot c$

POLYNOMIAL IDENTITIES

$(a + b)(c + d) = ac + ad + bc + bd$
(a.k.a. **FOIL**)

$(x + a)(x + b) = x^2 + (a + b)x + ab$
(a.k.a. **FOIL** with x-terms)

$(a + b)^2 = a^2 + 2ab + b^2$
(a.k.a. **square of a sum**)

$a^2 - b^2 = (a + b)(a - b)$
(a.k.a. **difference of squares**)

$a^3 - b^3 = (a - b)(a^2 + ab + b^2)$
(a.k.a. **difference of cubes**)

$a^3 + b^3 = (a + b)(a^2 - ab + b^2)$
(a.k.a. **sum of cubes**)

What is $\sqrt{-9}$? In other words, what number squared equals −9? If you square 3, the result is 9, and if you square −3, the result is also 9, because a negative multiplied by a negative is a positive. It seems like $\sqrt{-9}$ doesn't exist...but it does, in mathematics, as an imaginary number.

> The **imaginary number** i is defined as the square root of −1, so $i^2 = -1$.

Using imaginary numbers will allow you to solve problems that you've never before been able to solve. Now we can solve for $\sqrt{-9}$, which can be rewritten as $\sqrt{9 \cdot (-1)}$, or $\sqrt{9} \cdot \sqrt{-1}$, which equals $3i$.

> A **complex number** is a number in the form $a + bi$,
> where a and b are both real numbers.

This means that a complex number includes both a real-number component (a) and an imaginary-number component (bi). The number $-5 + 2i$ is a complex number. The imaginary number $3i$, which we mentioned above, is also a complex number—just one where the value of a in $a + bi$ is 0.

All real numbers are also complex numbers, just with $b = 0$ in the form $a + bi$.

The set of complex numbers includes all real numbers, all imaginary numbers, and all sums of real and imaginary numbers.

We can graphically represent complex numbers as points or vectors on a special complex number plane, which looks kind of like a real number plane (standard (x, y) coordinate grid) but with real numbers on the horizontal axis and imaginary numbers on the vertical axis. The point representing a complex number is plotted in alignment with its real-number component on the horizontal axis and its imaginary-number component on the vertical axis. The vector representing the same complex number is the vector from the origin to that point.

EXAMPLE 1

Show $2 + 4i$, $5 - i$, and $-3 + 3i$ on a complex number plane.

The complex number $2 + 4i$ consists of a positive 2 and a positive $4i$, so we'll draw a vector from the origin to the point that aligns with 2 on the real-number axis and $4i$ on the imaginary-number axis.

The complex number $5 - i$ aligns with 5 on the real-number axis and $-i$ on the imaginary-number axis.

The complex number $-3 + 3i$ aligns with -3 on the real-number axis and $3i$ on the imaginary-number axis.

The vectors and points representing $2 + 4i$, $5 - i$, and $-3 + 3i$ are shown on the complex number plane below.

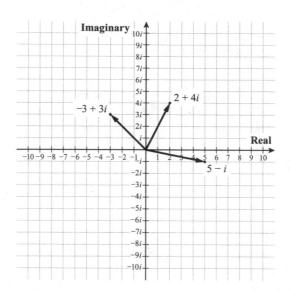

We can add complex numbers on the complex number plane by combining vectors. Start from the point of the first complex-number addend, and use the direction and length of the second complex number vector to find the point representing the sum.

Here is how you may see complex numbers on the ACT.

In the complex numbers, where $i^2 = -1$, which of the following is equal to the result of squaring the expression $(i + 5)$?

A. $5i$ $-1 + 10i + 25$

B. $25i$

C. $10i + 24$ $24 + 10i$

D. $5i - 25$

E. $i + 25$

What is the sum of 2 + 4i and 5 − i?

These vectors are shown in Example 1. Place the vector representing 5 − i, or the direct path that travels 5 units to the right and 1 unit down, starting at the end of the vector representing 2 + 4i.

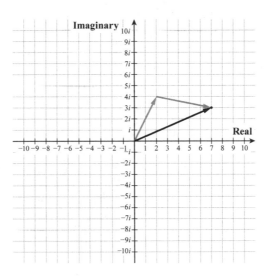

The sum is the point where the second vector ends, or 7 + 3i. The vector 7 + 3i, which moves 7 units to the right and 3 units up, is also shown on the graph.

We can also solve for the sum algebraically, as we would with real numbers. The **commutative**, **associative**, and **distributive properties** apply to all complex numbers.

$2 + 4i + 5 − i$
$2 + 5 + 4i − i$ Commutative property
$(2 + 5) + (4i − i)$ Associative property
$(2 + 5) + (4 − 1)(i)$ Distributive property
$7 + 3i$

EXAMPLE

Find the product of and simplify the expression below.
$(i - 4)(2i + 9)$

$2i^2 + 9i - 8i - 36$	Expand using FOIL.
$2i^2 + i - 36$	Combine like terms, $9i - 8i$.
$2(-1) + i - 36$	Substitute -1 for i^2.
$-2 + i - 36$	Multiply $2(-1)$.
$i - 38$	Combine like terms, $-2 - 36$, using the commutative property of addition.

EXAMPLE 4

Fully simplify the expression below.
$(8 + 10i)[12 - 2i - (7 - 6i)]$

$(8 + 10i)[12 - 2i - 7 + 6i]$	Distribute the subtraction through $(7 - 6i)$.
$(8 + 10i)(5 + 4i)$	Combine like terms using the commutative, associative, and distributive properties.
$8 \cdot 5 + 8 \cdot 4i + 10i \cdot 5 + 10i \cdot 4i$	Expand using FOIL.
$40 + 32i + 50i + 40i^2$	Perform all multiplication.
$40 + 82i + 40(-1)$	Combine like terms and substitute -1 for i^2.
$40 + 82i - 40$	Multiply.
$82i$	Combine like terms.

The expression $(8 + 10i)[12 - 2i - (7 - 6i)]$ is equal to $82i$.

Translate the words into an algebraic expression then expand and simplify.

$(i + 5)^2 = (i + 5)(i + 5)$	Squaring the expression means multiplying it by itself.
$= i^2 + 5i + 5i + 25$	Expand using the distributive property (FOIL).
$= i^2 + 10i + 25$	Combine like terms, $5i + 5i$.
$= -1 + 10i + 25$	Substitute -1 for i^2.
$= 10i + 24$	Combine like terms, $-1 + 25$.

The correct answer is (C).

Sometimes, the result of operations involving complex numbers is a fraction that has a complex number in the denominator. As you may recall from working with square roots in previous courses, the fraction is not fully simplified until the denominator is rational. Having an imaginary number in the denominator is anything *but* rational.

In order to simplify the denominator, we multiply both the numerator and denominator by the **conjugate** of the denominator. For example, to put $\dfrac{2}{3+6i}$ in standard form, we multiply both the numerator and denominator by $(3 - 6i)$. We do this because it produces a difference of squares in the denominator, including i^2, which is equal to the rational number -1. Once we replace i^2 with -1, the denominator becomes a real number, and the entire rational expression can be rewritten in the standard form for a complex number, $a + bi$.

Multiplying by the conjugate of the denominator is the same method you would use to simplify a fraction that contains the sum of a rational number and a square root in its denominator, such as $\dfrac{2}{3 + \sqrt{6}}$.

Simplify the expression $\dfrac{32}{3-\sqrt{7}i}$.

This fraction has an imaginary number, as well as a square root, in the denominator. To rewrite it in standard form, we must multiply both the numerator and the denominator by the conjugate of $3 - \sqrt{7}i$, which is $3 + \sqrt{7}i$.

$$\frac{32}{3-\sqrt{7}i} \cdot \frac{3+\sqrt{7}i}{3+\sqrt{7}i}$$

$$\frac{32\left(3+\sqrt{7}i\right)}{\left(3-\sqrt{7}i\right)\left(3+\sqrt{7}i\right)}$$ Multiply the numerators and the denominators.

$$\frac{32\left(3+\sqrt{7}i\right)}{3^2-\left(\sqrt{7}i\right)^2}$$ Rewrite the denominator as a difference of squares.

$$\frac{32\left(3+\sqrt{7}i\right)}{9-7i^2}$$ Find each squared term.

$$\frac{32\left(3+\sqrt{7}i\right)}{9-7(-1)}$$ Substitute -1 for i^2.

$$\frac{32\left(3+\sqrt{7}i\right)}{16}$$ Simplify the denominator.

$$2(3+\sqrt{7}i)$$ Simplify the fraction by dividing the numerator by the denominator, 16.

$$6+2\sqrt{7}i$$ Distribute to multiply through.

🔓 5

Lesson 1.2
Operations with Polynomials

Polynomials are terms or expressions that have one or more variables (raised to any whole-number exponent) and numbers. **Monomials** have one term, **binomials** have two, and **trinomials** have three.

VOLUME OF A CYLINDER: $V = \pi r^2 h$

VOLUME OF A SPHERE: $V = 4/3\, \pi r^3$

SOME RULES OF EXPONENTS: $a^0 = 1$
$a^m \cdot a^n = a^{m+n}$
$(a^m)^n = a^{mn}$

As with complex numbers, the commutative, associative, and distributive properties apply to polynomials. Use these properties, as well as the properties of exponents, to perform operations on polynomials. Always fully simplify your answer.

EXAMPLE

Expand and simplify: $(x - 5)(x^2 + 3x - 4)$

We can use the distributive property to rewrite the product as the sum of the product of x and the trinomial and the product of -5 and the trinomial.

$x(x^2 + 3x - 4) - 5(x^2 + 3x - 4)$

$x^3 + 3x^2 - 4x - 5x^2 - 15x + 20$ | Distribute the multiplication through to each term. | A subtraction of 5 is the same as an addition of -5. The negative sign needs to be included in the multiplication of each term in the trinomial: $(-5)(x^2) + (-5)(3x) + (-5)(-4)$.

$x^3 + 3x^2 - 5x^2 - 4x - 15x + 20$ | Use the commutative property to group like terms together.

$x^3 - 2x^2 - 19x + 20$ | Combine all like terms.

A real number can be described as a monomial in which the only x-term has an exponent of 0. Remember, a polynomial includes at least one variable raised to any whole-number exponent, and 0 is a whole number. The value of x^0 is 1, so the number 8 is the polynomial $8x^0$.

In this example, the product of two polynomials is another polynomial. In fact, whenever you add, subtract, or multiply polynomials, the result will always be another polynomial. We say that polynomials are closed under the operations of addition, subtraction, and multiplication. Even in the case of $(x + 6) + (-x + 2)$, the sum, 8, is still a polynomial. All real numbers are polynomials.

Integers, like polynomials, are closed under addition, subtraction, and multiplication. Notice that neither integers nor polynomials are closed under division. If you divide an integer by another integer, you may end up with a non-integer such as 3/4. If you divide a polynomial by another polynomial, you may end up with a non-polynomial such as $5/x$ or x^{-2}.

EXAMPLE

Uta has designed a vase that is cylindrical but has a solid glass hemispherical base, as shown below.

Remember, the volume of a cylinder is given by the formula $V = \pi r^2 h$, and the volume of a sphere is given by the formula $V = 4/3\, \pi r^3$.

Write a formula for the volume capacity of the vase, in terms of r, the radius of its circular base, and h, its height. Then, rewrite the formula according to a rule that the height of the vase must be 3 inches greater than the diameter of the vase.

The capacity of the vase is the total capacity of a cylinder this size ($\pi r^2 h$) minus the volume of the hemispherical base. A hemisphere is half of a sphere, so the volume of this hemisphere is $1/2\,(4/3\, \pi r^3)$, or $2/3\, \pi r^3$. Let's translate this information into an equation: The capacity of the vase is given by the formula $V = \pi r^2 h - 2/3\, \pi r^3$.

Here is how you may see operations with polynomials on the ACT.

The expression $(n^2 - 8n + 1)(n + 3)$ is equivalent to:

  A. $n^3 - 8n^2 - 24n + 3$
  B. $n^3 - 7n^2 - 21n + 3$
  C. $n^3 - 5n^2 + n + 3$
  D. $n^3 - 8n^2 + n + 3$
  E. $n^3 - 5n^2 - 23n + 3$

$n^3 - 5n^2 - 23n + 3$

If the height must be 3 inches greater than the diameter, then $h = d + 3$. The diameter is twice the length of the radius ($d = 2r$), so $h = 2r + 3$. Let's substitute this expression in terms of the radius for height in our volume formula.

$$V = \pi r^2(2r + 3) - 2/3\ \pi r^3$$

Expand and combine like terms to simplify.

$$V = 2\pi r^3 + 3\pi r^2 - 2/3\ \pi r^3$$
$$V = (2\pi r^3 - 2/3\ \pi r^3) + 3\pi r^2$$
$$V = \pi r^3(2 - 2/3) + 3\pi r^2$$
$$V = \pi r^3(4/3) + 3\pi r^2$$
$$V = 4/3\ \pi r^3 + 3\pi r^2$$

Keep an eye out for questions that ask for a specific unit of measurement. This formula gives volume in cubic inches, not capacity in ounces or liters, since h was defined in terms of inches.

We can also use our polynomial operation skills to uncover the pattern behind raising a binomial to various whole-number exponents.

EXAMPLE 8

Expand $(x + y)^1$, $(x + y)^2$, $(x + y)^3$, and $(x + y)^4$.

$(x + y)^1 = x + y$
$(x + y)^2 = (x + y)(x + y) = x(x + y) + y(x + y) = x^2 + xy + xy + y^2 = x^2 + 2xy + y^2$

Let's rewrite $(x + y)^3$ as $(x + y)(x + y)^2$ then substitute the expression we found for $(x + y)^2$.

$$\begin{aligned}(x + y)^3 &= (x + y)(x + y)^2 = (x + y)(x^2 + 2xy + y^2) \\ &= x(x^2 + 2xy + y^2) + y(x^2 + 2xy + y^2) \\ &= x^3 + 2x^2y + xy^2 + x^2y + 2xy^2 + y^3 \\ &= x^3 + 3x^2y + 3xy^2 + y^3\end{aligned}$$

Let's rewrite $(x + y)^4$ as $(x + y)(x + y)^3$ then substitute the expression we found for $(x + y)^3$.

$$\begin{aligned}(x + y)^4 &= (x + y)(x^3 + 3x^2y + 3xy^2 + y^3) \\ &= x(x^3 + 3x^2y + 3xy^2 + y^3) + y(x^3 + 3x^2y + 3xy^2 + y^3) \\ &= x^4 + 3x^3y + 3x^2y^2 + xy^3 + x^3y + 3x^2y^2 + 3xy^3 + y^4 \\ &= x^4 + 4x^3y + 6x^2y^2 + 4xy^3 + y^4\end{aligned}$$

Do you see a pattern? Each expanded expression begins with x raised to the power that $(x + y)$ was raised to, and each ends with y raised to that power. In each consecutive term within the expanded expression, the power of x decreases by 1 and the power of y increases by 1.

The coefficients of the terms also seem to follow a pattern: (1, 1), (1, 2, 1), (1, 3, 3, 1), and (1, 4, 6, 4, 1). Each set of coefficients is like a palindrome. Even more interesting, they follow the pattern of numbers shown in the triangle below. This pattern was discovered by Blaise Pascal and is now referred to as **Pascal's Triangle**.

The 1 in the first row represents $(x + y)^0$, which is equal to 1, because any non-zero number raised to a power of 0 is equal to 1. Then, as you saw from Example 8, the second row shows the coefficients for $(x + y)^1$, the third row shows the coefficients for $(x + y)^2$, and so on.

```
              1
           1     1
        1     2     1
     1     3     3     1
  1     4     6     4     1
1     5    10    10     5     1
```

The triangle continues infinitely beyond what is shown here, with an expanding base. Each number is the sum of the two numbers in the row above it immediately to the left and right of its placement.

The **Binomial Theorem** states this pattern formally with the equation

$$(x + y)^n = \binom{n}{0} x^n y^0 + \binom{n}{1} x^{n-1} y^1 + \binom{n}{2} x^{n-2} y^2 + \dots + \binom{n}{n-1} x^1 y^{n-1} + \binom{n}{n} x^0 y^n$$

where each $\binom{n}{k}$ is the binomial coefficient, as shown in Pascal's Triangle.

ACT A

We can use the distributive property to rewrite the product as the sum of the product of n and the trinomial and the product of 3 and the trinomial.

$n(n^2 - 8n + 1) + 3(n^2 - 8n + 1)$

$n^3 - 8n^2 + n + 3n^2 - 24n + 3$	Distribute the multiplication through to each term.
$n^3 - 8n^2 + 3n^2 + n - 24n + 3$	Use the commutative property to group like terms together.
$n^3 - 5n^2 - 23n + 3$	Combine all like terms.

The correct answer is (E).

EXAMPLE 9

Expand $(a^2 - 2ab)^5$.

First, rewrite the binomial as a sum, to fit the form for the Binomial Theorem: $[a^2 + (-2ab)]^5$.

The binomial is raised to a power of 5, so the coefficients we need are in the 6th row of Pascal's Triangle: 1, 5, 10, 10, 5, and 1.

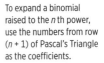

To expand a binomial raised to the nth power, use the numbers from row $(n + 1)$ of Pascal's Triangle as the coefficients.

$$[a^2 + (-2ab)]^5$$

$$= 1(a^2)^5(-2ab)^0 + 5(a^2)^4(-2ab)^1 +$$
$$10(a^2)^3(-2ab)^2 + 10(a^2)^2(-2ab)^3 +$$
$$5(a^2)^1(-2ab)^4 + 1(a^2)^0(-2ab)^5$$

$$= 1(a^{10})(1) + 5(a^8)(-2ab) + 10(a^6)(4a^2b^2) +$$
$$10(a^4)(-8a^3b^3) + 5(a^2)(16a^4b^4) + 1(1)(-32a^5b^5)$$

$$= a^{10} - 10a^9b + 40a^8b^2 - 80a^7b^3 + 80a^6b^4 - 32a^5b^5$$

Lesson 1.3
Polynomial Identities

Don't worry; there are other shortcuts for various polynomial operations that are much simpler than the Binomial Theorem! We call these shortcuts **polynomial identities**. A polynomial identity is any true equation that shows two equivalent polynomials, often revealing a special pattern that makes the identity more memorable.

To prove a polynomial identity, rewrite the expression on one side of the equation, usually by expanding, and show that it is equivalent to the expression on the other side of the equation.

EXAMPLE

Prove the polynomial identity $(x + y)^3 = x^3 + y^3 + 3xy(x + y)$.

$(x + y)^3 = (x + y)(x + y)^2$	Use the rule of exponents $a^{m+n} = a^m + a^n$.
$= (x + y)(x^2 + 2xy + y^2)$	Use the polynomial identity $(a + b)^2 = a^2 + 2ab + b^2$.
$= x(x^2 + 2xy + y^2) + y(x^2 + 2xy + y^2)$	Use the distributive property.
$= x^3 + 2x^2y + xy^2 + x^2y + 2xy^2 + y^3$	Use the distributive property.
$= x^3 + 3x^2y + 3xy^2 + y^3$	Combine like terms.
$= x^3 + y^3 + 3x^2y + 3xy^2$	Use the commutative property.
$= x^3 + y^3 + 3xy(x + y)$	Factor out $3xy$ from the last two terms.

Putting Polynomial Identities to Use

Polynomial identities are useful beyond simply providing an efficient way to expand basic polynomial products. For example, recognizing patterns from polynomial identities may allow you to solve for certain quantities without necessarily knowing the values of individual variables.

EXAMPLE

If $x + y = 8$ and $x^2 + y^2 = 40$, then $xy =$

A polynomial identity that includes both expressions $x + y$ and $x^2 + y^2$ is the square of a sum, $(x + y)^2 = x^2 + 2xy + y^2$.

$(x + y)^2 = x^2 + y^2 + 2xy$	Rearrange the terms to get $x^2 + y^2$ together.
$8^2 = 40 + 2xy$	Substitute 8 for $(x + y)$ and 40 for $(x^2 + y^2)$.
$64 = 40 + 2xy$	Square 8.
$24 = 2xy$	Subtract 40 from both sides.
$12 = xy$	Divide both sides by 2.

We were able to solve for the value of xy, even though we never specifically solved for the individual values of x and y.

The polynomial identity $(x^2 - y^2)^2 + (2xy)^2 = (x^2 + y^2)^2$, although not seen as often as those mentioned above, is useful for finding Pythagorean triples.

EXAMPLE 12

Find two Pythagorean triples using the polynomial identity above.

This polynomial identity is in the form $a^2 + b^2 = c^2$, with $a = x^2 - y^2$, $b = 2xy$, and $c = x^2 + y^2$. We need to choose a value of x and a value of y, then solve for a, b, and c. The absolute value of x must be greater than the absolute value of y, so that $x^2 - y^2$ will be positive. (The value of a, the length of one leg of the triangle, cannot be negative or zero.) Let's try $x = 2$ and $y = 1$.

$$a = x^2 - y^2 = 2^2 - 1^2 = 4 - 1 = 3$$
$$b = 2xy = 2(2)(1) = 4$$
$$c = x^2 + y^2 = 2^2 + 1^2 = 4 + 1 = 5$$

One Pythagorean triple is 3-4-5.

That was a very basic Pythagorean triple. Let's try some slightly larger numbers further apart (although you could certainly use 3 and 1, or 3 and 2, if you wanted). Let $x = 5$ and $y = 2$.

$$a = x^2 - y^2 = 5^2 - 2^2 = 25 - 4 = 21$$
$$b = 2xy = 2(5)(2) = 20$$
$$c = x^2 + y^2 = 5^2 + 2^2 = 25 + 4 = 29$$

Our second Pythagorean triple is 20-21-29.

Polynomial identities are especially useful for factoring polynomials. Factoring is the key to solving many problems involving polynomial equations.

EXAMPLE 13

Ethan threw a ball upward out the window of his apartment. The height of the ball in feet, h, is given by the equation $h = -16t^2 + 48t + 64$, where t is the time in seconds since Ethan threw the ball. After how many seconds did the ball hit the ground? After how many seconds did the ball reach its maximum height? What was the maximum height that the ball reached? Draw a graph showing the relationship between time and height.

When the ball hit the ground, $h = 0$, so we must solve the equation $0 = -16t^2 + 48t + 64$ for t.

$0 = -16(t^2 - 3t - 4)$	Factor out -16.
$0 = -16(t + 1)(t - 4)$	Factor the trinomial.
$0 = (t + 1)(t - 4)$	Divide both sides by -16.
$0 = t + 1 \qquad 0 = t - 4$	Set each factor equal to 0.
$t = -1 \qquad t = 4$	Solve for t.

The solution $t = -1$ does not make sense for this situation. The value of t must be positive, as it represents the amount of time since Ethan threw the ball. So, the answer is $t = 4$. The ball hit the ground 4 seconds after Ethan threw it.

The graph of a quadratic function is a parabola. To find the maximum, the turning point of the parabola, we must put the equation $h = -16t^2 + 48t + 64$ in vertex form. To do that, we'll complete the square.

$$h = -16(t^2 - 3t) + 64$$

Add the square of half the coefficient of t, 3. You must add the same amount to both sides of the equation, so don't forget to multiply by -16 on the left side.

$$
\begin{aligned}
h + (-16)(3/2)^2 &= -16[t^2 - 3t + (3/2)^2] + 64 \\
h + (-16)(9/4) &= -16(t - 3/2)^2 + 64 \\
h - 36 &= -16(t - 3/2)^2 + 64 \\
h &= -16(t - 3/2)^2 + 100
\end{aligned}
$$

The vertex is at (3/2, 100), or (1.5, 100). In other words, the ball reached its maximum height of 100 feet 1.5 seconds after Ethan threw it.

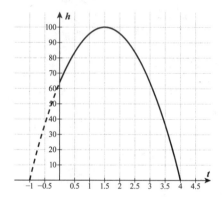

Lesson 1.4
Graphing Polynomials (Beyond Quadratics)

The vertex form of a quadratic function, $f(x) = a(x - h)^2 + k$, tells you that its graph is the result of shifting the graph of $f(x) = ax^2$ a total of h units to the right and k units up. If h is negative, the graph is shifted to the left, and if k is negative, the graph is shifted down.

A negative value of h will appear as an addition symbol in the vertex form of the equation, because subtraction of a negative is an addition of a positive. For example, the function $f(x) = (x + 3)^2$ written in vertex form is $f(x) = (x - (-3))^2$, so the value of h is -3.

> For any kind of function, the graph of $f(x) + k$ is the graph of $f(x)$ translated vertically—up if k is positive and down if k is negative—because the function value is adjusted at every point by exactly k units.

When a graph is shifted without changing the shape, orientation, or magnitude of the graph, we say that it is **translated**.

For example, the function $f(x) = x^4 - 2x^2 + 10$ will always be exactly 10 units greater than the function $f(x) = x^4 - 2x^2$ for any given x-value, so its graph is identical but shifted 10 units up.

> The graph of $f(x - h)$, for any kind of function, is the graph of $f(x)$ translated h units to the right.

If you subtract a constant, h, from x, then the function value is what it would be for the x-value that is h units to the left in the parent function. In other words, the graph is shifted to the right when the subtracted h is positive. For example, $g(x) = (x - 6)^4$ is $f(x) = x^4$ shifted 6 units to the right. The value of $g(8)$ is the same as the value of $f(2)$, 16, as shown in the graphs on the following page.

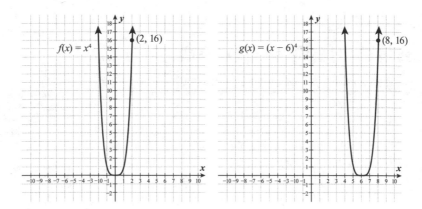

EXAMPLE 14

The graph of $f(x) = x^3$ is shown in the standard (x, y) coordinate plane below. Write an equation that describes the graph of the cubic function shifted 2 units to the right and 5 units down.

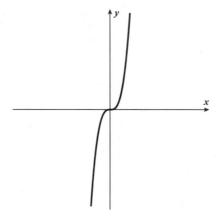

The function graph will be shifted 2 units to the right, so we must subtract 2 from x inside the cubed expression. The function graph will be shifted 5 units down, so we must subtract 5 from the function value (outside of the cubed expression). The translated function has the equation $f(x) = (x - 2)^3 - 5$.

To check our answer, let's test the equation in relation to the graph. The new function equation should shift each point 2 units right and 5 units down. Let's sketch the translated graph next to the graph of $f(x) = x^3$.

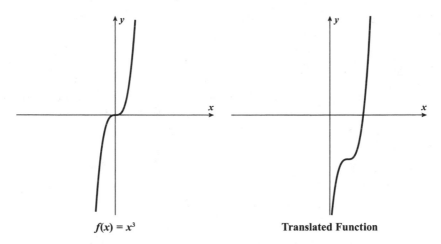

$f(x) = x^3$ **Translated Function**

When $x = 2$, the original function $f(x) = x^3$ has a value of 8, so the graph of $f(x) = x^3$ passes through the point (2, 8). Let's test the equation $f(x) = (x - 2)^3 - 5$ to see if it produces the point that is on the translated function graph, 2 units to the right and 5 units below (2, 8), which is (4, 3). Test $x = 4$ in $f(x) = (x - 2)^3 - 5$.

$$f(4) = (4 - 2)^3 - 5 = 2^3 - 5 = 8 - 5 = 3$$

The function correctly produced the point (4, 3). If you choose any point on $f(x) = x^3$, the function $f(x) = (x - 2)^3 - 5$ produces a point 2 units right and 5 units down from there.

Now you know how function graphs of degree n are translated when written in the form $y = (x - h)^n + k$, where h and k are constants. However, you do not need to write a polynomial in this form to graph it. Solving for the zeros of a polynomial function allows you to make a rough sketch of its graph, using its x-intercepts and end behavior.

> The **roots** of a single-variable polynomial expression are the solutions for the variable when the polynomial is set equal to 0. These roots are called **zeros** of the corresponding polynomial function. The zeros are the x-intercepts of the graph of the function.

In Example 13 (the ball-throw example), the roots of the quadratic are −1 and 4, because these are the values of t that are solutions to $-16t^2 + 48t + 64 = 0$. The values −1 and 4 are the zeros of the function $h(t) = -16t^2 + 48t + 64$. The graph of the function crosses the x-axis at −1 and 4.

The **end behavior** of a function describes how it behaves for very small (approaching negative infinity) and very large (approaching infinity) values of x. Because of the relationship of exponents, the value of the term with the greatest exponent overpowers any other terms within the expression, regardless of their coefficients. For example, for very large or very small values of x, $y = x^4 - 100x^3$ will have a greater x^4 value than a $100x^3$ value, and the ends of the graph will follow the rules for x^4.

When a polynomial function is of an even degree, the x-values at either end raised to that power produce very large positive numbers. When the leading coefficient for a polynomial function of even degree is positive, then both ends have very large positive values, and both arms of the graph point up. When the leading coefficient for a polynomial function of even degree is negative, this negative number multiplies by the large positive value of x raised to that power to create a negative value approaching negative infinity, and both arms of the graph point down.

When a polynomial function is of an odd degree, such as with a leading term of x^3 or x^5, then the sign of x is preserved for extreme values of x (a large negative x-value raised to an odd power produces a very large negative function value), and the arms of the function point in opposite directions from one another.

End Behavior of Polynomial Functions

A polynomial function of even degree with a positive leading coefficient has both arms pointing up.

A polynomial function of even degree with a negative leading coefficient has both arms pointing down.

A polynomial function of odd degree with a positive leading coefficient has its left arm pointing down and its right arm pointing up.

A polynomial function of odd degree with a negative leading coefficient has its left arm pointing up and its right arm pointing down.

EXAMPLE 15 🔒

Sketch the function $y = -3x^3 - 9x^2 + 30x$.

First, factor the polynomial $-3x^3 - 9x^2 + 30x$, set equal to 0.

$-3x(x^2 + 3x - 10) = 0$ -3 and x are each a common factor in all three terms.

$-3x(x + 5)(x - 2) = 0$ Factor the trinomial $x^2 + 3x - 10$.

Solve for each factor set equal to 0.

$-3x = 0$	$x + 5 = 0$	$x - 2 = 0$
$x = 0$	$x = -5$	$x = 2$

The zeros are -5, 0, and 2, so these are the x-intercepts of the function. The function is a cubic with a negative coefficient of x^3, so the left end of the graph points up and the right end points down. Let's sketch the graph using this information.

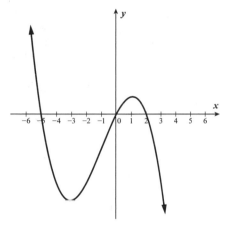

To find more points on the graph, you would use the given function equation to solve for y-values produced by various x-values.

This is just a rough sketch. To graph more precisely, we would need to find at least a few more points on the graph.

Here is how you may see equations of polynomial graphs on the SAT.

The graph of the function $f(x)$ contains the points $(-6, 0)$, $(3/2, 0)$, and $(1, 0)$. Which of the following could be f?

    (A) $f(x) = 2x^3 + 7x^2 - 27x + 18$

    B) $f(x) = 2x^3 + 13x^2 + 3x - 18$

    C) $f(x) = 2x^3 - 17x^2 + 33x - 18$

    D) $f(x) = 2x^3 - 7x^2 - 27x - 18$

Notice that the graph in Example 15 has two **turning points**, one of which is a local minimum and the other of which is a local maximum. A cubic may have two turning points, as in this case. It also may have no turning points, as in the graph of $y = x^3$, shown in Example 14. A quadratic has exactly one turning point, which we call the vertex, as you learned in Algebra I. Straight lines have no turning points, as they never turn but continue forever in one direction.

> A polynomial of degree n has at most $(n - 1)$ turning points, which are local maximums or minimums.

With an understanding of how polynomial functions are graphed, we can also work in the other direction. Given the zeros of a polynomial function and a little more information about its magnitude, we can write the equation of the function.

EXAMPLE 16

What polynomial function is shown on the graph below?

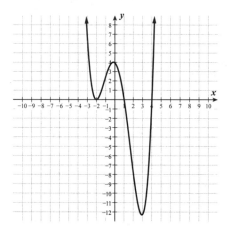

The factor $[x - (-2)]$ is the same as $(x + 2)$.

The three x-intercepts of this graph are -2, 1, and 4, so these are the zeros of the function. We can write each factor in the form $(x - a)$, where a is a zero of the function. So, $(x + 2)$, $(x - 1)$, and $(x - 4)$ are all factors of the polynomial. Let's try multiplying them together.

$$(x + 2)(x - 1)(x - 4)$$
$$(x + 2)(x^2 - 5x + 4)$$
$$x^3 - 3x^2 - 6x + 8$$

Both ends of the graph point upward, so it must be a polynomial function of an even degree. However, the polynomial $x^3 - 3x^2 - 6x + 8$ is of degree 3—an odd degree, not even. How can that be? Well, when an x-intercept is also a local vertex, or turning point, then the zero is a **repeated zero**. In this case, -2 is a repeated zero, so we must use it twice when writing the polynomial.

$(x + 2)(x + 2)(x - 1)(x - 4)$ Write the polynomial using all four zeros: -2, -2, 1, and 4.

$(x + 2)(x^3 - 3x^2 - 6x + 8)$ Substitute the expression we found for $(x + 2)(x - 1)(x - 4)$.

$x^4 - x^3 - 12x^2 - 4x + 16$ Expand and write in standard form.

If the polynomial function graphed were $y = x^4 - x^3 - 12x^2 - 4x + 16$, then it would have a y-intercept of 16 (the solution for y when $x = 0$). Instead, this graph shows a y-intercept of 4. If we multiply our polynomial by 1/4, that will adjust the magnitude, without affecting the zeros, to produce the correct y-intercept.

$$y = 1/4 \, (x^4 - x^3 - 12x^2 - 4x + 16)$$
$$y = 1/4 \, x^4 - 1/4 \, x^3 - 3x^2 - x + 4$$

The polynomial function shown in the graph is $y = 1/4 \, x^4 - 1/4 \, x^3 - 3x^2 - x + 4$ in standard form, or $y = 1/4 \, (x + 2)(x + 2)(x - 1)(x - 4)$ in factored form.

Some polynomials have fewer real zeros than their highest degree for other reasons. If the graph reaches a turning point before intersecting with the x-axis (a local maximum below the x-axis or a local minimum above the x-axis), then the polynomial will have two fewer real zeros.

The three given points all have an $f(x)$-value of 0, which means that they all represent x-intercepts of the function. So, the zeros of $f(x)$ are -6, 3/2, and 1. We can write each factor of $f(x)$ in the form $(x - a)$, where a is a zero of the function.

$(x - (-6))(x - 3/2)(x - 1)$
$(x + 6)(x - 3/2)(x - 1)$

However, the multiplication of these factors will be a little messy with a fraction involved. Let's write a different binomial factor to represent the zero 3/2. The only rule is that the binomial factor, when set equal to 0, must solve as $x = 3/2$. The binomial $(2x - 3)$ fits this description.

$(x + 6)(2x - 3)(x - 1)$
$(x + 6)(2x^2 - 5x + 3)$ Use FOIL to multiply $(2x - 3)(x - 1)$.
$x(2x^2 - 5x + 3) + 6(2x^2 - 5x + 3)$ Use the distributive property.
$2x^3 - 5x^2 + 3x + 12x^2 - 30x + 18$ Distribute through each multiplication.
$2x^3 + 7x^2 - 27x + 18$ Combine all like terms.

A function with zeros -6, 3/2, and 1 could have other zeros, as well (which would result in additional binomial factors), or a different magnitude (which would mean multiplying this polynomial by a constant). However, we do not have to worry about these possibilities, because $f(x) = 2x^3 + 7x^2 - 27x + 18$ is one of the given answer choices. The correct answer is (A).

Lesson 1.5
Tips for Factoring Polynomials

QUADRATIC FORMULA:

$$x = \frac{-b \pm \sqrt{b^2 - 4ac}}{2a}$$

for a quadratic equation of the form $ax^2 + bx + c = 0$

Much of the art of factoring is in recognizing familiar patterns. For example, a good way to start is by looking for a common factor in all terms. Next, check to see if the polynomial matches any of the common polynomial identities.

In some cases, only part of a given polynomial may need to be factored.

EXAMPLE 17

Which of the following is equivalent to $x^{12} - 14x^6 + 52$?

A. $(x^6 - \sqrt{7}x^3)^2 + 52$

B. $(x^4 - 7x^2)^3 + 52$

C. $(x^6 - 7)^2 + 3$

D. $(x^2 - 7)^6 + 3$

The exponents are very large, which makes this appear difficult to rewrite, but it's actually fairly simple once you see the relationship between the exponents. Notice that x^{12} is the same as $(x^6)^2$. This allows us to rewrite the expression in terms of x^6, as $(x^6)^2 - 14x^6 + 52$. We can even let $a = x^6$, which makes the expression $a^2 - 14a + 52$. Now it seems more manageable!

There are no integer factors of 52 that add up to −14, so we can't factor the entire expression using FOIL. However, we don't need to factor the entire polynomial. The answer choices show a factorization of only part of the polynomial. Notice that they each include a squared binomial.

We can use $a^2 - 14a$ to write a squared binomial, with the number 52 broken up to accommodate that. The polynomial identity for the square of a difference is $(a - b)^2 = a^2 - 2ab + b^2$. If we view $a^2 - 14a$ as $a^2 - 2ab$, then $14 = 2b$ and $b = 7$. So, $(a - 7)^2 = a^2 - 14a + 49$. Rewrite the original expression to include this trinomial.

This method is called completing the square, and it's the same method we use to rewrite quadratic functions in vertex form (see Example 13).

$a^2 - 14a + 52 = a^2 - 14a + 49 + 3$	Rewrite the expression, showing 52 as the sum of 49 and 3.
$= (a - 7)^2 + 3$	Substitute $(a - 7)^2$ for $a^2 - 14a + 49$.
$= (x^6 - 7)^2 + 3$	Substitute x^6 for a.

The correct answer is (C).

To check our work, let's expand the expression $(x^6 - 7)^2 + 3$ to make sure it is equivalent to $x^{12} - 14x^6 + 52$.

$(x^6 - 7)^2 + 3 = (x^6)^2 - 2(x^6)(7) + 7^2 + 3$	Expand the squared binomial using the polynomial identity $(a - b)^2 = a^2 - 2ab + b^2$.
$= x^{12} - 14x^6 + 49 + 3$	Evaluate exponents and products.
$= x^{12} - 14x^6 + 52$	Combine like terms.

You could also expand using FOIL, just in case you have misremembered the identity used here.

Although you may sometimes be expected to factor only part of a polynomial, as in Example 17, you will usually need to factor the entire polynomial, as when solving for zeros of a function.

The act of factoring a polynomial is essentially dividing the polynomial by one factor and writing the quotient as the other factor. In order for this to work, the factor must divide evenly into the original polynomial, with no remainder.

> The **Remainder Theorem** states that when a polynomial $p(x)$ is divided by $(x - a)$, the remainder is $p(a)$, so $p(a)$ is equal to 0 if and only if $(x - a)$ is a factor of $p(x)$.

This can help us confirm a root if we're having trouble factoring a polynomial.

In Example 15, if we didn't see the common factor of $-3x$, we might have felt stuck in our efforts to factor the cubic $-3x^3 - 9x^2 + 30x$. But, in that case, we could have tested various values as our guesses for possible zeros, using the Remainder Theorem. For example, let's test 0, 1, and 2.

$$p(x) = -3x^3 - 9x^2 + 30x$$
$$p(0) = -3(0^3) - 9(0^2) + 30(0) = 0 + 0 + 0 = 0$$

Because $p(0) = 0$, we know that $(x - 0)$, or x, is a factor of $-3x^3 - 9x^2 + 30x$, according to the Remainder Theorem.

$$p(1) = -3(1^3) - 9(1^2) + 30(1) = -3 - 9 + 30 = 18$$

Because $p(1)$ is not equal to 0, we know that $(x - 1)$ is not a factor.

$$p(2) = -3(2^3) - 9(2^2) + 30(2) = -24 - 36 + 60 = 0$$

Because $p(2) = 0$, we know that $(x - 2)$ is a factor.

So far, we know that x and $(x - 2)$ are both factors of $-3x^3 - 9x^2 + 30x$. If we also recognize that -3 can be factored out, we can work out the following:

$$
\begin{aligned}
-3x^3 - 9x^2 + 30x &= -3(x^3 + 3x^2 - 10x) &&\text{Factor out } -3. \\
&= -3x(x^2 + 3x - 10) &&\text{Factor out } x. \\
&= -3x(x - 2)(\underline{}) &&\text{Factor out } (x - 2).
\end{aligned}
$$

At this point, we should be able to recognize that the other factor is $(x + 5)$. However, even if we still don't see it, we can fall back on the good old **quadratic formula**.

$$x = \frac{-3 \pm \sqrt{3^2 - 4(1)(-10)}}{2(1)} = \frac{-3 \pm \sqrt{49}}{2} = \frac{-3 \pm 7}{2}$$

$$x = \frac{-3 + 7}{2} = \frac{4}{2} = 2 \qquad x = \frac{-3 - 7}{2} = -\frac{10}{2} = -5$$

We had already found the root 2, so this work with the quadratic formula gave us the other root, −5. The full factorization of the polynomial is $-3x(x - 2)(x + 5)$.

Almost any time you are expected to factor a higher-degree polynomial on a math test, the roots will be rational numbers.

The Remainder Theorem provides us with a way to test potential roots in the given polynomial, but sometimes we need help with the first step of figuring out what roots to test. The Rational Root Theorem provides a set of potential roots, narrowing down our choices from any possible rational numbers to a smaller number of possibilities.

> The **Rational Root Theorem** states that if a polynomial has any rational roots, they must be in the form $\pm \dfrac{\text{factor of the constant term}}{\text{factor of the leading coefficient}}$.

If we were trying to find the roots of $x^4 - x^3 - 12x^2 - 4x + 16$, we would know that any rational roots would be a positive or negative ratio of a factor of 16 (the constant term) to a factor of 1 (the coefficient of x^4). Factors of 16 include 1, 2, 4, 8, and 16. Factors of 1 include...just 1. So, any rational roots of $x^4 - x^3 - 12x^2 - 4x + 16$ must be in the set $\{\pm 1/1, \pm 2/1, \pm 4/1, \pm 8/1, \pm 16/1\}$, or $\{-16, -8, -4, -2, -1, 1, 2, 4, 8, 16\}$. Indeed, the roots are −2 (repeated), 1, and 4, as you saw in Example 16.

EXAMPLE 18

Factor $2x^3 - x^2 - 7x + 6$.

Let's start by using the Rational Root Theorem. Any rational roots of this polynomial will be in the set of $\pm \dfrac{\text{factors of 6}}{\text{factors of 2}}$, or $\pm \dfrac{\{1, 2, 3, 6\}}{\{1, 2\}}$. These possible roots include $\pm\{1, 2, 3, 6, 1/2, 3/2\}$.

Notice that 2/2 simplifies to 1, which is the same as 1/1, and 6/2 simplifies to 3, which is the same as 3/1. That's why our set of 8 potential roots turned out to be only 6 distinct potential roots.

We can use the Remainder Theorem to test any of these roots. Let's test 1.

$$p(x) = 2x^3 - x^2 - 7x + 6$$

$$p(1) = 2(1^3) - 1^2 - 7(1) + 6 = 2 - 1 - 7 + 6 = 0$$

Because $p(1) = 0$, $(x - 1)$ is one factor of the polynomial.

Because 2 is the leading coefficient and is not a common factor of all terms, it's likely that one of the roots is $\pm 1/2$ or $\pm 3/2$. Let's use the Remainder Theorem again.

$$p(1/2) \quad = 2(1/2)^3 - (1/2)^2 - 7(1/2) + 6$$
$$= 1/4 - 1/4 - 7/2 + 6$$
$$= 5/2 \qquad\qquad \text{Not a root}$$

$$p(-1/2) \quad = 2(-1/2)^3 - (-1/2)^2 - 7(-1/2) + 6$$
$$= -1/4 - 1/4 + 7/2 + 6$$
$$= 9 \qquad\qquad \text{Not a root}$$

$$p(3/2) \quad = 2(3/2)^3 - (3/2)^2 - 7(3/2) + 6$$
$$= 27/4 - 9/4 - 21/2 + 6$$
$$= 9/2 - 21/2 + 6$$
$$= 0$$

Sure, you could write the factor $(x - 3/2)$ based on the root 3/2, but then you'd have to later multiply by 2. The factor $(2x - 3)$ also corresponds to a root of 3/2 but doesn't include any fractions to complicate things.

Aha! 3/2 is a root, so $(2x - 3)$ is one factor of the polynomial. So far we have $(x - 1)(2x - 3)$, which expands to $2x^2 - 5x + 3$. We need one more factor to complete $2x^3 - x^2 - 7x + 6$. At this point, we can just divide as we would with numbers, in long-division form. We know that the polynomial $2x^2 - 5x + 3$ must be multiplied by x to get a beginning of $2x^3$.

$$
\begin{array}{r}
x \\
2x^2 - 5x + 3 \overline{)\ 2x^3 - x^2 - 7x + 6} \\
\underline{-(2x^3 - 5x^2 + 3x)} \\
4x^2 - 10x + 6
\end{array}
$$

The difference between $2x^3 - x^2 - 7x + 6$ and $2x^3 - 5x^2 + 3x$ is $4x^2 - 10x + 6$. It looks like we should now multiply $2x^2 - 5x + 3$ by 2.

$$
\begin{array}{r}
x + 2 \\
2x^2 - 5x + 3 \overline{)\ 2x^3 - x^2 - 7x + 6} \\
\underline{-(2x^3 - 5x^2 + 3x)} \\
4x^2 - 10x + 6 \\
\underline{-(4x^2 - 10x + 6)} \\
0
\end{array}
$$

We need to factor the polynomial to find the zeros of the function.

$-1/2\ (x^5 - 7x^3 - 18x)$	Factor out $-1/2$.
$-1/2\ x(x^4 - 7x^2 - 18)$	Factor out the common factor, x.
$-1/2\ x[(x^2)^2 - 7(x^2) - 18]$	See the trinomial in the form $a^2 - 7a - 18$.
$-1/2\ x(x^2 - 9)(x^2 + 2)$	Factor the trinomial as $(a - 9)(a + 2)$.
$-1/2\ x(x + 3)(x - 3)(x^2 + 2)$	Factor the difference of squares, $x^2 - 9$.

The expression $x^2 + 2$ cannot be further factored using real numbers, and it does not correspond to any x-intercepts, or real zeros, of the function. Set each of the other factors equal to 0 to solve for the real zeros of $f(x)$.

$$-1/2\ x = 0 \qquad x + 3 = 0 \qquad x - 3 = 0$$
$$x = 0 \qquad\qquad x = -3 \qquad\quad x = 3$$

The real zeros of $f(x)$ are -3, 0, and 3. The correct answer is (C).

There is no remainder from our division, so $(x + 2)(2x^2 - 5x + 3) = 2x^3 - x^2 - 7x + 6$. The polynomial $2x^3 - x^2 - 7x + 6 = (x - 1)(2x - 3)(x + 2)$, fully factored.

Alternatively, we could have continued testing other potential roots using the Remainder Theorem. We eventually would have found that $p(-2) = 2(-2)^3 - (-2)^2 - 7(-2) + 6 = -16 - 4 + 14 + 6 = 0$, which would have told us that $(x + 2)$ was a factor.

Polynomials with Complex Roots

What about a polynomial like $x^2 + 9$? According to the Rational Root Theorem, any rational roots would be in the set $\pm\{1, 3, 9\}$, but all six of those numbers fail the Remainder Theorem test. For example, $(-3)^2 + 9 = 9 + 9 = 18$, not 0. None of these six potential rational roots is a root of $x^2 + 9$. The graph of $y = x^2 + 9$ never crosses the x-axis, and the function has no real zeros.

Even though there are no real roots of $x^2 + 9$, there are imaginary roots! Remember i? If $x = 3i$, then $x^2 + 9 = 0$, because $(3i)^2 + 9 = 9i^2 + 9 = 9(-1) + 9 = -9 + 9 = 0$. Actually, there is one more imaginary-number solution to this equation. To solve $x^2 + 9 = 0$, we should factor $x^2 + 9$ as a difference of squares, $x^2 - (-9)$, where -9 is the value of $(3i)^2$. So, $(x + 3i)(x - 3i) = 0$, and $-3i$ and $3i$ are the complex-number solutions for x.

A polynomial with real coefficients will only have complex number roots in conjugate pairs, for complex numbers that include imaginary components. In other words, if one root of a polynomial with real coefficients is of the form $a + bi$, with $b \neq 0$, then $a - bi$ is also a root of the polynomial.

Other polynomial identities can also be applied to complex number situations, as long as we are careful to remember that $i^2 = -1$.

EXAMPLE

Find all solutions to the equation $2x^4 - 30x^2 = 32$.

$2x^4 - 30x^2 - 32 = 0$	Move all terms to one side.
$2(x^4 - 15x^2 - 16) = 0$	Factor out 2.
$2[(x^2)^2 - 15(x^2) - 16] = 0$	Recognize that the trinomial is in the form $a^2 - 15a - 16$.
$2(x^2 - 16)(x^2 + 1) = 0$	Factor the trinomial as $(a - 16)$ $(a + 1)$.
$2(x + 4)(x - 4)(x + i)(x - i)$	Factor differences of squares, both real and imaginary.
$x = -4, x = 4, x = -i, x = i$	Solve each factor set equal to 0.

The solutions to the equation $2x^4 - 30x^2 = 32$ are $x = \pm 4$ and $x = \pm i$.

For example, the square of a binomial identity, $(a + b)^2 = a^2 + 2ab + b^2$, can be applied to complex numbers: $(a + bi)^2 = a^2 + 2abi + b^2i^2$, so $(a + bi)^2 = a^2 + 2abi - b^2$. Also, FOIL becomes $(a + bi)(c + di) = ac + (ad + bc)i - bd$. The fact that $i^2 = -1$ produces a negative final term in both cases.

If we were only looking for the real solutions to this equation, we could graph the function $y = 2x^4 - 30x^2 - 32$ and find its x-intercepts, where $y = 0$. For this function, we will need to adjust the scale on our graphing calculator or program, to see how the function behaves below the x-axis. The graph of $y = 2x^4 - 30x^2 - 32$ is shown below.

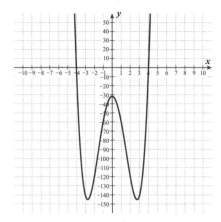

The only two x-intercepts are at -4 and 4, and neither is at a turning point, so -4 and 4 are the only two real solutions to the given equation. With three turning points and both ends pointing up, the graph must represent a polynomial of at least degree 4, which indicates that there are additional solutions, but the graph does not indicate what those additional solutions are. To find the imaginary-number solutions, we need to solve the equation algebraically, as shown on the previous page.

EXAMPLE 🔒20

Find all solutions to the equation $3x^2 - 6x = -5$.

We can rewrite this equation as $3x^2 - 6x + 5 = 0$. However, it does not seem to be factorable, so we must use the quadratic formula.

If you graph this quadratic, either by hand or using graphing technology, you will see that the parabola does not cross the x-axis. That means that the equation has no real-number solutions. Both solutions must include imaginary numbers.

$$x = \frac{-(-6) \pm \sqrt{(-6)^2 - 4(3)(5)}}{2(3)}$$

$$= \frac{6 \pm \sqrt{36 - 60}}{6}$$

$$= \frac{6 \pm \sqrt{-24}}{6}$$

$$= \frac{6 \pm \sqrt{4} \cdot \sqrt{-1} \cdot \sqrt{6}}{6}$$

$$= \frac{6 \pm 2i\sqrt{6}}{6}$$

$$= \frac{3 \pm i\sqrt{6}}{3}$$

The solutions to $3x^2 - 6x = -5$ are $x = \dfrac{3 + i\sqrt{6}}{3}$ and $x = \dfrac{3 - i\sqrt{6}}{3}$. These roots can also

be written as $1 + \dfrac{\sqrt{6}}{3} i$ and $1 - \dfrac{\sqrt{6}}{3} i$, in the standard form for complex numbers.

Lesson 1.6
Polynomials in the Real World

Everything you've learned about polynomials can help you in real-world applications.

EXAMPLE

A company manufactures solar chargers. They have discovered that the demand curve for this particular solar charger is described by the equation $q = -1/10\, p^2 + 490$, where q is the number of chargers sold in a month at a price of p dollars. It costs the company $10 per charger to manufacture them. Write and graph a polynomial function representing the company's monthly profit, t, in terms of p. Using graphing technology, find the solar charger price that will maximize the company's profits.

Profit is equal to total revenue minus costs. The company's monthly revenue from selling the solar chargers is found by multiplying the number of solar chargers sold by the price per charger. In this case, the number sold is q, or $(-1/10\, p^2 + 490)$, and the price per charger is p. The company's revenue for the month is $(-1/10\, p^2 + 490)(p)$ dollars.

The company's costs for manufacturing q solar chargers at $10 per charger is $10q$ dollars, or $10(-1/10\, p^2 + 490)$ dollars.

So, the company's monthly profit from the solar chargers is given by the equation $t = (-1/10\, p^2 + 490)(p) - 10(-1/10\, p^2 + 490)$. Expand and simplify this equation to put it in standard form.

$$t = -1/10\, p^3 + 490p + p^2 - 4900$$
$$t = -1/10\, p^3 + p^2 + 490p - 4900$$

In order for the polynomial to be in standard form, the terms must be arranged in order of decreasing powers of p.

The function is a cubic with a negative leading coefficient, so the left arm points up and the right arm points down. The y-intercept (or t-intercept) is −4900. To find the x-intercepts, we need to factor the polynomial and solve for its zeros. To factor, look back at the function's original form, $t = (-1/10\ p^2 + 490)(p) - 10(-1/10\ p^2 + 490)$. The binomial is a common factor in the two terms.

$t = (-1/10\ p^2 + 490)(p - 10)$ Factor out the binomial.
$t = -1/10\ (p^2 - 4900)(p - 10)$ Factor out −1/10.
$t = -1/10\ (p + 70)(p - 70)(p - 10)$ Factor the difference of squares.

If $0 = -1/10\ (p + 70)(p - 70)(p - 10)$, then $p = -70$, $p = 70$, or $p = 10$.

The function t has zeros of −70, 10, and 70, so these are the x-intercepts (p-intercepts) of its graph. The function generally looks like this.

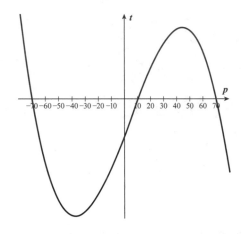

However, negative values of p do not make sense in the context of the problem. (The company can't charge a negative amount for each charger. That would mean paying people to take them!) Also, p cannot be greater than 70, because no solar chargers get sold at prices of $70 or more. Our graph should only show the curve for $0 \le p \le 70$.

Let's use the factored form of the equation to find more points on the graph.

$t = -1/10\ (p + 70)(p - 70)(p - 10)$
When $p = 20$, $t = -1/10\ (20 + 70)(20 - 70)(20 - 10) = 4500$
When $p = 30$, $t = -1/10\ (30 + 70)(30 - 70)(30 - 10) = 8000$
When $p = 40$, $t = -1/10\ (40 + 70)(40 - 70)(40 - 10) = 9900$
When $p = 50$, $t = -1/10\ (50 + 70)(50 - 70)(50 - 10) = 9600$
When $p = 60$, $t = -1/10\ (60 + 70)(60 - 70)(60 - 10) = 6500$

The graph passes through the points (20, 4500), (30, 8000), (40, 9900), (50, 9600), and (60, 6500), in addition to the intercepts (0, −4900), (10, 0), and (70, 0).

Here is a better graph of the function t.

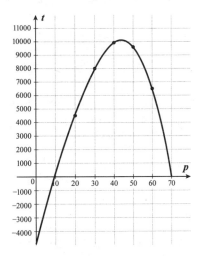

You can also use a graphing calculator or graphing program to graph the function $t = -1/10\, p^3 + p^2 + 490p - 4900$. Zoom in to locate the maximum on the curve. It appears to be at (44, 10,077.6). This represents the point of the greatest possible profit for the company. To maximize their profits, t, they should set the price, p, of the solar charger at $44.

CHAPTER 1 PRACTICE QUESTIONS

Directions: Complete the following open-ended problems as specified by each question stem. For extra practice after answering each question, try using an alternative method to solve the problem or check your work.

1. Find all of the zeros of the polynomial function $f(x) = x^4 - x^2 - 20$.

$(x^2 - 5)(x^2 + 4)$

$(x - 2i)(x + \sqrt{5})(x - \sqrt{5})(x + 2i)$

2. Find all the zeros of $f(x) = x^4 - 3x^3 + 6x^2 + 2x - 60$ given that $1 + 3i$ is a zero of f.

$(x + 2)(x - 3)(1 + 3i)(1 - 3i)$

3. Find the solutions of the quadratic $3x^2 - 2x + 5 = 0$.

$\dfrac{1 \pm i\sqrt{14}}{3}$

4. Write $6i(5 - 2i)$ in the form $a + bi$.

$12 + 30i$

5. Expand and simplify $(2 + 3i)^2 + (2 - 3i)^2$.

$-5 + 12i + 13 - 12i = 8$

6. Expand the binomial $(2t - s)^5$ using Pascal's Triangle to determine the coefficients.

$32t^5 - 80t^4s + 80t^3s^2 - 40t^2s^3 + 10ts^4 + s^5$

7. Given $f(x) = x^3 - x^2 - 2x$, find a new function, g, which is created by shifting $f(x)$ 4 units to the right.

$g(x) = (x - 4)(x - 6)(x - 3)$

8. A group of students are painting a poster to promote the math team. They paint a grid on a piece of poster board, when a centipede starts walking on the board. They step away and watch as the centipede trails the fresh paint as it walks, forming a path that can be described by the cubic function $f(x) = x^3 + 6x^2 - x - 30$. At what points will the centipede cross the horizontal axis and bring with it more paint to continue its odd drawing?

$(-5, 0)(-3, 0)(2, 0)$

9. A few friends have recently started a technology company and want to have the best possible launch for their new smartphone. The demand for the phone follows a curve as described by the equation $d = \dfrac{-1}{20}c^2 + 125$, where c is the price at which each phone is being sold and d is the number of phones that would be sold per month at that price. Due to cost of labor and technology, however, it costs \$15 to manufacture each phone. Write and graph a polynomial function representing the company's monthly profit, p, in terms of c. At what price point would the company start to make money by selling the phones? At what price point after that would the company begin to lose money?

$p = \dfrac{-1}{20}c^3 + 125c + \dfrac{3}{4}c^2 - 1875$

$0 = -\dfrac{1}{20}c^3 + \dfrac{3}{4}c^2 + 125c - 1875$

$= -\dfrac{1}{20}(c^3 - 15c^2 - 2500c + 37500)$

$= -\dfrac{1}{20}(c - 15)(c^2 - 2500)$

$= -\dfrac{1}{20}(c - 15)(c + 50)(c - 50)$

$\boxed{< \$15 \ \& \ > \$50}$

$f(x) = x(x^2 - x - 2)$
$= x(x - 2)(x + 1)$

$x = \dfrac{2 \pm \sqrt{4 - 60}}{6}$

$= \dfrac{2 \pm 2i\sqrt{14}}{6}$

$2 \mid 1 \quad -3 \quad 6 \quad 2 \quad -60$
$ \quad -2 \quad 10 \quad -32 \quad 60$
$3 \mid 1 \quad -5 \quad 16 \quad -30 \quad 0$
$ \quad 3 \quad -6 \quad 30$
$1 \quad -2 \quad 10 \quad 0$

$x = \dfrac{2 \pm \sqrt{4 - 40}}{2}$
$= 1 \pm 3i$

SOLUTIONS TO CHAPTER 1 PRACTICE QUESTIONS

1. **$x = \pm 2i$ and $x = \pm \sqrt{5}$**

   To begin, you should recognize that the trinomial $x^4 - x^2 - 20$ is in the form $(x^2)^2 - (x^2) - 20$. You can then set up parentheses and input the first terms:

   $$(x^2 \quad)(x^2 \quad)$$

   Next, identify factors of 20 that will add together to get a sum of -1 to arrive at the middle term. The factors of 20 are 1 and 20, 2 and 10, and 4 and 5. The only possible pairing of numbers are 4 and 5, so place those inside the parentheses.

   $$(x^2 \quad 4)(x^2 \quad 5)$$

   Since the middle term of the polynomial is negative, the larger number, 5, must be negative and since the final term is negative, the 4 must be positive.

   $$(x^2 + 4)(x^2 - 5)$$

   To find the zeros, set the factored expression equal to zero and solve for each x-value.

   $(x^2 + 4)(x^2 - 5) = 0$
   $x^2 + 4 = 0 \qquad x^2 - 5 = 0$
   $x^2 = -4 \qquad\; x^2 = 5$
   $x = \pm 2i \qquad x = \pm \sqrt{5}$

2. **$1 + 3i$, $1 - 3i$, 3, and -2**

   When a complex zero is provided, we need to remember that complex zeros occur in conjugate pairs. What that means is since $1 + 3i$ is a zero, $1 - 3i$ must also be a zero. This means that both

   $$[x \quad (1 + 3i)] \quad \text{and} \quad [x - (1 - 3i)]$$

   are factors of $f(x)$. Multiplying these two factors produces

   $$[x - (1 + 3i)][x - (1 - 3i)] = [(x - 1) - 3i][(x - 1) + 3i]$$
   $$= (x - 1)^2 - 9i^2$$
   $$= x^2 - 2x + 1 - 9(-1)$$
   $$= x^2 - 2x + 10$$

   Using long division, we can divide $x^2 - 2x + 10$ into $f(x)$ to obtain the following.

   $$
   \begin{array}{r}
   x^2 - x - 6 \\
   x^2 - 2x + 10 \overline{)\, x^4 - 3x^3 + 6x^2 + 2x - 60} \\
   \underline{-(x^4 - 2x^3 + 10x^2)} \\
   -x^3 - 4x^2 + 2x \\
   \underline{-(-x^3 + 2x^2 - 10x)} \\
   -6x^2 + 12x - 60 \\
   \underline{-(-6x^2 + 12x - 60)} \\
   0
   \end{array}
   $$

Therefore, we have

$f(x) = (x^2 - 2x + 10)(x^2 - x - 6)$
$\qquad = (x^2 - 2x + 10)(x - 3)(x + 2)$

and we conclude that the zeros of f are

$$1 + 3i, \ 1 - 3i, \ 3, \text{ and } -2.$$

3. $\dfrac{1}{3} + \dfrac{\sqrt{14}}{3}i$ **and** $\dfrac{1}{3} - \dfrac{\sqrt{14}}{3}i.$

First, try to determine whether or not the provided quadratic is easy to factor. The quadratic does not look easy to factor, so you should utilize the quadratic formula with $a = 3$, $b = -2$, and $c = 5$.

$$x = \frac{-(-2) \pm \sqrt{(-2)^2 - 4(3)(5)}}{2(3)}$$

$$= \frac{2 \pm \sqrt{-56}}{6}$$

$$= \frac{2 \pm 2\sqrt{14}i}{6}$$

$$= \frac{1}{3} \pm \frac{\sqrt{14}}{3}i$$

Thus, the given equation has two solutions:

$$\frac{1}{3} + \frac{\sqrt{14}}{3}i \qquad \text{and} \qquad \frac{1}{3} - \frac{\sqrt{14}}{3}i.$$

4. **$12 + 30i$**

When combining concepts such as the distribution property and imaginary numbers, pay close attention to the signs at each step of work. Begin by distributing $6i$ to each term in the parentheses.

$$6i(5 - 2i) = 6i(5) + 6i(-2i)$$
$$= 30i - 12i^2$$
$$= 30i - 12(-1)$$
$$= 30i + 12$$

Therefore, the result is $12 + 30i$.

5. **-10**

Proceed with caution here. You can't just square the individual terms inside the parentheses. First, to expand, FOIL out the individual monomials or use the identity $(a + b)^2 = a^2 + 2ab + b^2$:

$(2 + 3i)^2 = (2 + 3i)(2 + 3i)$ $\qquad$ $(2 - 3i)^2 = (2 - 3i)(2 - 3i)$
$\qquad = 4 + 6i + 6i + 9i^2$ $\qquad\qquad\quad$ $= 4 - 6i - 6i + 9i^2$
$\qquad = 4 + 12i + 9(-1)$ $\qquad\qquad\quad$ $= 4 - 12i + 9(-1)$
$\qquad = 4 + 12i - 9$ $\qquad\qquad\qquad\quad$ $= 4 - 12i - 9$
$\qquad = -5 + 12i$ $\qquad\qquad\qquad\qquad$ $= -5 - 12i$

Next, combine the like terms through addition.

$$(-5 + 12i) + (-5 - 12i) = -5 + 12i - 5 - 12i = -5 - 5 = -10.$$

6. $32t^5 - 80t^4s + 80t^3s^2 - 40t^2s^3 + 10ts^4 - s^5$

Recall that Pascal's Triangle is a handy tool to determine quickly what the coefficients of each term will be when expanded out, with the top row accounting for the term $(x + y)^0$, which is equal to 1. The Binomial Theorem also accounts for the shortcut of figuring the exponents for each term: The first term's exponents decrease from left to right, from 5 to 0 in this case, whereas the second term's exponents increase from left to right, from 0 to 5. First, determine the coefficients from Pascal's Triangle. The binomial is raised to a power of 5, so the 6th row of the triangle is needed: 1, 5, 10, 10, 5, and 1. Next, using the coefficients in conjunction with the Binomial Theorem, begin expanding out the binomial.

$$(2t - s)^5 = 1(2t)^5(-s)^0 + 5(2t)^4(-s)^1 + 10(2t)^3(-s)^2 + 10(2t)^2(-s)^3 + 5(2t)^1(-s)^4 + 1(2t)^0(-s)^5$$

Carefully apply the exponents to each term and pay close attention to the signs.

$$1(2t)^5(-s)^0 + 5(2t)^4(-s)^1 + 10(2t)^3(-s)^2 + 10(2t)^2(-s)^3 + 5(2t)^1(-s)^4 + 1(2t)^0(-s)^5$$
$$= 1(32t^5)(1) + 5(16t^4)(-s) + 10(8t^3)(s^2) + 10(4t^2)(-s^3) + 5(2t)(s^4) + 1(1)(-s^5)$$
$$= 32t^5 - 80t^4s + 80t^3s^2 - 40t^2s^3 + 10ts^4 - s^5.$$

7. $x^3 - 13x^2 + 54x - 72$

Recall that when shifting polynomial expressions horizontally, the change occurs inside the parentheses affecting the x term, so the appropriate substitution here is to evaluate $f(x - 4)$ and combine like terms.

$$f(x - 4) = (x - 4)^3 - (x - 4)^2 - 2(x - 4)$$

First begin by expanding $(x - 4)^3$.

$$\begin{aligned}(x - 4)^3 &= (x - 4)(x - 4)^2 \\ &= (x - 4)(x^2 - 8x + 16) \\ &= x(x^2 - 8x + 16) - 4(x^2 - 8x + 16) \\ &= x^3 - 8x^2 + 16x - 4x^2 + 32x - 64 \\ &= x^3 - 12x^2 + 48x - 64\end{aligned}$$

Next $(x - 4)^2$ needs to be expanded, but through the work in the previous step, the result of $x^2 - 8x + 16$ was already gathered, so begin combining all like terms to simplify and arrive at the function g.

$$\begin{aligned}f(x - 4) &= (x^3 - 12x^2 + 48x - 64) - (x^2 - 8x + 16) - 2(x - 4) \\ &= x^3 - 12x^2 + 48x - 64 - x^2 + 8x - 16 - 2x + 8 \\ &= x^3 - 13x^2 + 54x - 72\end{aligned}$$

So $g(x) = x^3 - 13x^2 + 54x - 72$.

Alternatively, you could factor the given cubic, $x^3 - x^2 - 2x$, to find the zeros of $f(x)$. It factors as $x(x - 2)(x + 1)$, so the graph of $f(x)$ intersects the x-axis at 0, 2, and −1. If the graph is shifted 4 units to the right, then the new graph has x-intercepts 4, 6, and 3. Using these zeros, you can write the equation of this new function: $g(x) = (x - 4)(x - 6)(x - 3)$. If you expand, this becomes $g(x) = x^3 - 13x^2 + 54x - 72$.

8. **$x = -5, -3,$ and 2**

This question is asking for the x-intercepts, or the zeros, of $f(x)$, so you must find the roots of the cubic. Use the Rational Root Theorem first to determine possible roots. In this case, the coefficient of the x^3 term is 1, so the possible roots are the positive or negative factors of 30, giving options of $\pm 1, \pm 2, \pm 3, \pm 5, \pm 6, \pm 10, \pm 15,$ and ± 30. Try testing the roots in the original function using the Remainder Theorem:

$f(x) = (1)^3 + 6(1)^2 - 1 - 30 = 1 + 6 - 1 - 30 = -24$, so 1 is not a root.
$f(x) = (-1)^3 + 6(-1)^2 - (-1) - 30 = -1 + 6 + 1 - 30 = -24$, so −1 is not a root.
$f(x) = (2)^3 + 6(2)^2 - (2) - 30 = 8 + 24 - 2 - 30 = 0$, so 2 is a root and therefore $(x - 2)$ is a factor. Divide that out to see what remains:

$$
\begin{array}{r}
x^2 + 8x + 15 \\
x - 2 \overline{)\, x^3 + 6x^2 - x - 30} \\
-(x^3 - 2x^2) \\
\hline
8x^2 - x \\
-(8x^2 - 16x) \\
\hline
15x - 30 \\
-(15x - 30) \\
\hline
0
\end{array}
$$

The remaining quadratic is $x^2 + 8x + 15$. This can be factored to $(x + 3)(x + 5)$, so the three binomial factors of the function are $(x - 2)(x + 3)(x + 5)$, and the three points at which the centipede will cross the horizontal axis are at $x = -5, -3,$ and 2.

9. **At a price point above \$15, the company makes money; at a price point above \$50, they lose money again.**

First, since profit is the same thing as revenue minus cost, both of those need to be combined to

form the desired equation. The revenue is given by the product of the number of phones

sold and the price per phone; since the number sold is d and the price for each phone is c,

the revenue would be $\left(\dfrac{-1}{20}c^2 + 125\right)(c)$. The cost to make d phones would be $15d$, or 15

$\left(\dfrac{-1}{20}c^2 + 125\right)$. Given this information, the company's profit from phone sales would be given

by $p = \left(\dfrac{-1}{20}c^2 + 125\right)(c) - 15\left(\dfrac{-1}{20}c^2 + 125\right)$. Expand this and simplify where possible to put this

in standard form:

$$p = \frac{-1}{20}c^3 + 125c + \frac{3}{4}c^2 - 1875$$

$$p = \frac{-1}{20}c^3 + \frac{3}{4}c^2 + 125c - 1875$$

This cubic equation has a negative leading coefficient, so it will have a left arm facing up and a right arm facing down. Its y-intercept, when $c = 0$, is -1875.

The company would break even when they make $0 profit, so factor the polynomial and find the zeros of the function. Going back to the original form rather than standard form, one of the factors is pulled out.

$$p = \left(\frac{-1}{20}c^2 + 125 \right)(c) - 15\left(\frac{-1}{20}c^2 + 125 \right)$$

$$p = (c - 15)\left(\frac{-1}{20}c^2 + 125 \right)$$

The fractional part of the second piece can be pulled out:

$$p = -\frac{1}{20}(c - 15)(c^2 - 2500)$$

This gives the difference of two squares for the other piece:

$$p = -\frac{1}{20}(c - 15)(c - 50)(c + 50)$$

This gives three zeros, at $c = -50$, 15, and 50. A rough sketch of this information can be seen here:

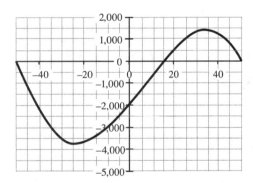

Because it is impossible to sell phones at a negative price in the real world, we can ignore the x-intercept at -50. The point at which the company starts making money is the x-intercept where the profits change from negative to positive. That would be $15, so they would begin to make money if the phones were priced above $15. Also, if they priced their phones at more than $50 per phone, the company would begin to lose money again, as at that point the function is about to dip back below the x-axis.

REFLECT

Congratulations on completing Chapter 1!
Here's what we just covered.
Rate your confidence in your ability to:

- Perform arithmetic operations
with complex numbers
① ② ③ ④ ⑤

- Perform arithmetic operations
with polynomials
① ② ③ ④ ⑤

- Use polynomial identities to
solve for certain quantities, find
Pythagorean triples, and factor
polynomials
① ② ③ ④ ⑤

- Graph polynomial functions and
derive polynomial equations from
given graphs
① ② ③ ④ ⑤

- Factor higher-degree polynomials
completely, into any real- and/or
imaginary-number roots
① ② ③ ④ ⑤

- Create and interpret polynomial
functions and their graphs to
represent and explain real-world
relationships
① ② ③ ④ ⑤

**If you rated any of these topics lower than you'd like, consider reviewing
the corresponding lesson before moving on, especially if you found yourself
unable to correctly answer one of the related end-of-chapter questions.**

Access your online student tools for a
handy, printable list of Key Points for this
chapter. These can be helpful for retaining
what you've learned as you continue to
explore these topics.

Chapter 2
Systems of Equations and Rational Expressions

2

GOALS **By the end of this chapter, you will be able to**

- Determine how many roots a polynomial has, based on the degree of the polynomial

- Factor polynomials with imaginary number coefficients

- Create and solve systems of equations to solve single-variable polynomial equations

- Identify and apply direct, inverse, joint, and combined variations

- Perform arithmetic operations with rational expressions

- Rewrite rational expressions as the sum of a polynomial and a remainder expression

Lesson 2.1
The Fundamental Theorem of Algebra

In Chapter 1, we saw that single-variable polynomials with real-number coefficients can have real roots, imaginary roots, or both.

> The **Fundamental Theorem of Algebra** asserts that any single-variable polynomial of degree n with complex coefficients has exactly n roots, including all instances of repeated roots.

Every single-variable polynomial of degree 1 or greater with complex coefficients has at least one complex root.

A single-variable polynomial of degree 1 or greater is a polynomial that includes one variable, such as x, whether to the first power (such as in $7x - 1$) or to a higher power (such as in $-2x^2$ or $x^9 + 6x^4$).

The single-variable polynomial $1/4\, x^4 - 1/4\, x^3 - 3x^2 - x + 4$ is of degree 4 (the highest power of x in the polynomial), and it has exactly 4 roots (-2, -2, 1, and 4), even though -2 is a repeated root. This is a case where all coefficients are real and all roots are real. Real numbers are a subset of complex numbers, so the Fundamental Theorem of Algebra applies in these cases, but it also applies when coefficients and/or roots include imaginary numbers. $2x^4 - 30x^2 - 32$, another polynomial of degree 4, has exactly 4 roots (-4, 4, $-i$, and i), even though two of them are imaginary.

EXAMPLE 1

How many roots does $3z^2 + 4z + 5$ have? What are the roots?

This is a polynomial of degree 2, so it has two roots, according to the Fundamental Theorem of Algebra.

The polynomial does not seem factorable, so we'll use the quadratic formula,

$$z = \frac{-b \pm \sqrt{b^2 - 4ac}}{2a} .$$

$$z = \frac{-4 \pm \sqrt{4^2 - 4(3)(5)}}{2(3)}$$

$$= \frac{-4 \pm \sqrt{-44}}{6}$$

$$= \frac{-4 \pm 2i\sqrt{11}}{6}$$

$$= \frac{-2 \pm i\sqrt{11}}{3}$$

The two roots can also be written in standard form as $-2/3 + \frac{\sqrt{11}}{3}i$ and $-2/3 - \frac{\sqrt{11}}{3}i$.

The standard form of a complex number is $a + bi$, where a and b are real.

You can use the quadratic formula to find the roots of any polynomial of degree 2, which will always produce two roots (because of the ± sign), even if they are the same root (in cases of root multiplicity) or complex roots (as in Example 1). This echoes the Fundamental Theorem of Algebra.

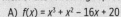

Here is how you may see the Fundamental Theorem of Algebra on the SAT.

Which of the following represents a polynomial function with roots of {-2, -2, and 5}?

A) $f(x) = x^3 + x^2 - 16x + 20$

B) $f(x) = x^3 - x^2 - 16x - 20$

C) $f(x) = x^2 + 3x - 10$

D) $f(x) = x^2 - 3x - 10$

$(x+2)^2(x-5)$

$(x^2+4x+4)(x-5)$

$(x^3-x^2-16x-20)$

So far, we have only looked at real polynomials. A **real polynomial** is a polynomial with real coefficients for all of its terms.

An **irreducible quadratic** is a real quadratic that has no real roots, but even so, it still has two complex roots, as in Example 1. An irreducible quadratic cannot be factored unless you use imaginary numbers in the factors.

> Every real polynomial can be factored into a product of real linear factors and real irreducible quadratic factors.

EXAGINE 2

Billy noticed that the polynomials he factored for homework always factored into real linear factors, with at most one real irreducible quadratic. He proposed a theory that every real polynomial has at most two roots that are not real. Graciela disproved his theory with the counterexample of $2x^4 + 7x^2 + 3$. Show how her example disproves Billy's theory.

The polynomial $2x^4 + 7x^2 + 3$ is a real polynomial, because the coefficients 2, 7, and 3 are real numbers. To factor this polynomial, recognize that the trinomial is in the form $2a^2 + 7a + 3$, which factors as $(2a + 1)(a + 3)$.

$$2(x^2)^2 + 7(x^2) + 3 = (2x^2 + 1)(x^2 + 3)$$

Both factors are irreducible quadratics. The roots of $2x^2 + 1$ are $\dfrac{i\sqrt{2}}{2}$ and $-\dfrac{i\sqrt{2}}{2}$.

The roots of $x^2 + 3$ are $i\sqrt{3}$ and $-i\sqrt{3}$. So, all four roots of $2x^4 + 7x^2 + 3$ are imaginary

numbers. Graciela's example is a real polynomial that has more than two roots that are

not real.

The Fundamental Theorem of Algebra covers polynomials with "complex coefficients," so it applies even when one or more of the coefficients is not real.

EXAMPLE 3

How many roots does the polynomial $x^3 - 2ix^2 + 15x$ have? What are the roots?

This is a polynomial of degree 3, so it has three roots.

$x^3 - 2ix^2 + 15x = x(x^2 - 2ix + 15)$ Factor out the common factor, x.

The x term has an i in its coefficient, so we should rewrite the constant term, 15, in terms of i.

$$x[x^2 - 2ix - (-15)]$$
$$x(x^2 - 2ix - 15i^2)$$ Substitute i^2 for -1.
$$x(x + 3i)(x - 5i)$$ Factor the trinomial.

The three roots of $x^3 - 2ix^2 + 15x$ are 0, $-3i$, and $5i$.

Alternatively, after factoring out x to get $x(x^2 - 2ix + 15)$, we could use the quadratic formula to solve for the roots of $x^2 - 2ix + 15$.

> If the coefficient of x^2 or the constant in a quadratic contains an imaginary number, the quadratic formula may result in a very ugly pair of expressions, including $\sqrt{i}$. We will not deal with those situations here, but know that any resulting solutions (roots) can still be written in the form $a + bi$, the standard form of a complex number, where a and b are real.

$$x = \frac{-(-2i) \pm \sqrt{(-2i)^2 - 4(1)(15)}}{2(1)}$$

$$= \frac{2i \pm \sqrt{4i^2 - 60}}{2}$$

$$= \frac{2i \pm \sqrt{4(-1) - 60}}{2}$$

$$= \frac{2i \pm \sqrt{-64}}{2}$$

$$= \frac{2i \pm 8i}{2}$$

$$x = \frac{2i + 8i}{2} = 5i \qquad\qquad x = \frac{2i - 8i}{2} = -3i$$

The Fundamental Theorem of Algebra tells us that a single-variable polynomial with three roots is of degree three, or a cubic, even in the case that one root is repeated. You can eliminate (C) and (D).

Write the factor that corresponds to each given root: the factor that will equal 0 if x is equal to the given root. Then multiply all three factors.

$(x + 2)(x + 2)(x - 5)$
$(x + 2)(x^2 - 3x - 10)$ Use FOIL to multiply $(x + 2)(x - 5)$.
$x(x^2 - 3x - 10) + 2(x^2 - 3x - 10)$ Use the distributive property to rewrite as a sum of products.
$x^3 - 3x^2 - 10x + 2x^2 - 6x - 20$ Use the distributive property to multiply through.
$x^3 - x^2 - 16x - 20$ Combine all like terms.

The correct answer is (B).

Lesson 2.2
Systems of Equations

A **system of equations** consists of two or more equations. Any solution to a system of equations is a solution to each of the individual equations in the system.

One more way to solve a single-variable polynomial equation is by creating and solving a system of equations in two variables.

> A solution to a system of equations is a coordinate pair (x, y) that makes both equations true. Each equation in the system shows y equal to a polynomial in terms of x, so the x-value of a solution to the system represents a solution to the equation where the two polynomials of x terms are set equal to each other.

EXAMPLE 4

Solve the equation $1/2\, x^5 + x^4 - 3/2\, x^3 - x^2 = 5x + 6$.

Alternatively, you could rewrite the equation in any arrangement you like, such as $1/2\, x^5 + x^4 = 3/2\, x^3 + x^2 + 5x + 6$. This would produce a different system of equations, but the solutions for x would still be the same.

We can rewrite this as a system of equations by setting each side of the equation equal to y.

$$y = 1/2\, x^5 + x^4 - 3/2\, x^3 - x^2$$

$$y = 5x + 6$$

An x-value that produces the same y-value in each equation makes $1/2\, x^5 + x^4 - 3/2\, x^3 - x^2$ equal to $5x + 6$. In other words, it will be a solution to our original single-variable equation. Using graphing technology, we can graph $y = 1/2\, x^5 + x^4 - 3/2\, x^3 - x^2$ and $y = 5x + 6$, as shown below.

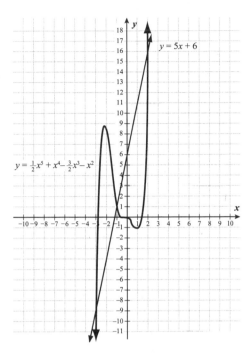

$y = 5x + 6$

$y = \frac{1}{2}x^5 + x^4 - \frac{3}{2}x^3 - x^2$

A graphing calculator allows you to be precise about points of intersection. However, when looking at a graph with unlabeled points, you cannot assume that the function graphs intersect exactly at (−3, −9). It could be (−2.9, −8.9), which is why it's important to check your solutions.

The points where the graphs intersect are the solutions to the system of equations: (−3, −9), (−1, 1), and (2, 16). Let's confirm these solutions using the equations.

When $x = -3$, $1/2\ (-3)^5 + (-3)^4 - 3/2\ (-3)^3 - (-3)^2 = -243/2 + 81 + 81/2 - 9 = -9$
   and $5(-3) + 6 = -15 + 6 = -9$.

When $x = -1$, $1/2\ (-1)^5 + (-1)^4 - 3/2\ (-1)^3 - (-1)^2 = -1/2 + 1 + 3/2 - 1 = 1$
   and $5(-1) + 6 = -5 + 6 = 1$.

When $x = 2$, $1/2\ (2^5) + 2^4 - 3/2\ (2^3) - 2^2 = 16 + 16 - 12 - 4 = 16$ and $5(2) + 6 = 10 + 6 = 16$.

The points (−3, −9), (−1, 1), and (2, 16) are indeed solutions to the system of equations, so their x-values (−3, −1, and 2) are solutions to the equation $1/2\ x^5 + x^4 - 3/2\ x^3 - x^2 = 5x + 6$.

Since this polynomial equation is of degree 5, it has a total of five solutions, but we have only found three so far. Let's move all terms to one side and solve for the remaining zeros.

$1/2\ x^5 + x^4 - 3/2\ x^3 - x^2 - 5x - 6 = 0$

Our solutions so far correspond to factors $(x + 3)$, $(x + 1)$, and $(x - 2)$, which multiply together to produce $x^3 + 2x^2 - 5x - 6$. Let's divide $1/2\ x^5 + x^4 - 3/2\ x^3 - x^2 - 5x - 6$ by $x^3 + 2x^2 - 5x - 6$ to see what factors remain.

See Example 18 in Chapter 1, Lesson 1.5, for a walk-through of how to perform long division of polynomials.

$$\frac{1}{2}x^2 + 1$$

$$x^3 + 2x^2 - 5x - 6\ \overline{\big)\ \tfrac{1}{2}x^5 + x^4 - \tfrac{3}{2}x^3 - x^2 - 5x - 6}$$

$$-\left(\tfrac{1}{2}x^5 + x^4 - \tfrac{5}{2}x^3 - 3x^2\right)$$

$$\qquad\qquad x^3 + 2x^2 - 5x - 6$$

$$\qquad\qquad -(x^3 + 2x^2 - 5x - 6)$$

$$\qquad\qquad\qquad\qquad\qquad 0$$

The other factor is $(1/2\ x^2 + 1)$. The roots of this quadratic are $\pm i\sqrt{2}$.

The five solutions to $1/2\ x^5 + x^4 - 3/2\ x^3 - x^2 = 5x + 6$ are $x = -3$, $x = -1$, $x = 2$, $x = -i\sqrt{2}$, and $x = i\sqrt{2}$.

Here is how you may see systems of equations on the SAT.

$$y = -2x + 3$$
$$y = 2x^2 - 12x + 11$$

$2x^2 - 10x + 8$

$2(x^2 - 5x + 4)$

How many values of (x, y) satisfy the system of equations above?

A) 0

B) 1

C) 2

D) Infinitely many

Lesson 2.3
Variation Equations

A **variation equation** can be written as a variable set equal to a single term in which a constant is multiplied and/or divided by one or more other variables.

In a **direct variation**, two variables are directly proportional; one is a constant multiple of the other. A direct variation is of the form $a = kb$, where a and b are variables and k is some constant, known as the **constant of variation**. Another way to view a direct variation is in the form $a/b = k$, meaning that the two variables have a constant ratio. If one variable increases, the other variable must increase proportionally (by the same factor).

In a direct variation $a = kb$, we say, "a varies directly as b." In more common, everyday language, we say, "a varies directly with b."

In an **inverse variation**, one variable is inversely proportional to the other variable. The product of the two variables is constant, so $ab = k$, or $a = k/b$, where a and b are variables and k is the constant of variation. If one variable increases, the other variable must decrease by the same factor, so that the product is still the constant, k.

In a **joint variation**, a variable varies directly with more than one other variable. A joint variation in three variables is in the form $a = kbc$, where a, b, and c are variables and k is the constant of variation. Here, a varies directly as the product of b and c. If the value of b or c is held constant, then a varies directly as the other variable.

In a **combined variation**, a variable both varies directly with one or more variables and varies inversely with one or more other variables. The simplest form of a combined variation is $a = k \cdot b/c$ (which can also be written as $a = kb/c$), where a, b, and c are variables and k is the constant of variation. Here a varies directly as b and varies inversely as c, when the other variable is held constant.

**Describe the variation relationship and the value of *k*
for each of the volume equations below.**

$$V_{\text{rectangular prism}} = lwh$$

$$V_{\text{pyramid}} = 1/3\ Bh$$

$$V_{\text{cylinder}} = \pi r^2 h$$

For a rectangular prism, volume varies jointly as *l*, *w*, and *h* (length, width, and height), and $k = 1$.

For a pyramid, volume varies jointly as *B* (the area of its base) and *h* (height), and $k = 1/3$.

For a cylinder, volume varies directly as *h*, when *r* is held constant, so $k = \pi r^2$ in that case. We cannot say that the volume varies jointly as *r* and *h*, because *r* is squared. The relationship between *V* and *r* is not directly proportional when *h* is held constant. (If *r* doubles, *V* quadruples, not doubles.)

To solve algebraically, substitute $-2x + 3$ for *y* in the second equation. After setting the entire thing equal to 0, by adding, subtracting, and dividing from both sides, you should be able to factor the quadratic to find that the *x*-values of 1 and 4 are solutions. These can then be substituted into each equation to solve for their corresponding *y*-values, which yields (1, 1) and (4, −5).

Alternatively, you could graph both given equations, as shown below, and look for the points of intersection. The correct answer in either method is (C).

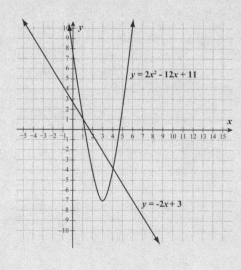

You can use variation relationships to set up equations and solve problems.

EXAMPLE 6

At 4:00 P.M., Jorge and Martin started driving toward one another on the same highway, each at a constant speed. They passed one another when Jorge had driven 80 miles and Martin had driven 96 miles. Write an equation that relates Jorge's speed to Martin's speed. If they passed one another at 5:20 P.M., at what speeds were they each driving?

For this situation, we can use the relationship distance = rate × time, or $d = rt$, and rewrite it as $t = d/r$. Let's call Jorge's rate j and Martin's rate m. For Jorge, $t = 80/j$, and for Martin, $t = 96/m$. Time is inversely proportional to speed. Since Jorge and Martin drove the same period of time until passing one another, t has the same value in both equations. We can substitute to get $80/j = 96/m$. Cross-multiply to get $80m = 96j$. If we solve for j, we get $j = 80/96\, m$, or $j = 5/6\, m$ simplified. This direct variation equation relates Jorge's speed to Martin's speed.

If they passed one another at 5:20 P.M., then they each drove for 1 hour and 20 minutes, or 1 1/3 hours. We can say $t = 4/3$, where t is in hours. Substitute this into Martin's $t = 96/m$ equation to get $4/3 = 96/m$. The solution is $m = 72$, so Martin's driving speed was 72 miles per hour.

The units must match, so we must express time in hours to match speed in miles per hour.

Jorge's speed in relation to Martin's speed is given by $j = 5/6\, m$, so $j = 5/6\,(72) = 60$. Jorge was driving at a speed of 60 miles per hour. We could also have used the equation $t = 80/j$ with $t = 4/3$ to solve for j, also resulting in $j = 60$.

The $d = rt$ relationship is a very simple one, but certain complicated relationships can still be understood in terms of variation relationships.

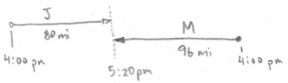

Caitlin plans to take out a business loan, the size of which will be determined by what monthly payment amount she can afford for a five-year repayment of the loan. She has found a bank that will give her a loan of any amount between $5,000 and $50,000 at a 3.6% interest rate compounded annually. To find the amount, P, of her monthly payment due, she uses the formula $P = \dfrac{rA}{1-\left(1+r\right)^{-n}}$, where A represents the amount of the loan, r represents the interest rate per month (as a decimal), and n represents the total number of monthly payments. What is the relationship between the loan amount and the monthly payment amount? Write and graph an equation representing this relationship. Use the graph to determine what loan amount Caitlin should apply for, if she can put $450 per month toward its repayment.

Here is how you may see combined variations on the ACT.

The number of minutes it takes a group of waiters at a certain catering company to set up for a dinner event varies directly with the square root of the number of guests and inversely with the number of waiters working the event. If c represents the constant of variation, which of the following expressions represents the number of minutes it will take w waiters to set up for a dinner event with n guests?

A. $cw\sqrt{n}$

B. $\dfrac{w\sqrt{n}}{c}$

C. $\dfrac{c}{w\sqrt{n}}$

D. $\dfrac{cw}{\sqrt{n}}$

E. $\dfrac{c\sqrt{n}}{w}$

$\dfrac{c\sqrt{n}}{w}$

Even though the formula $P = \dfrac{rA}{1-\left(1+r\right)^{-n}}$ looks messy, the values of r and n are constant for this particular situation. So, P varies directly as A, with a constant of variation of $\dfrac{r}{1-\left(1+r\right)^{-n}}$.

The 3.6% interest rate is an annual interest rate, compounded annually, so r, the interest rate per month, is 0.036/12, or 0.003. The variable n represents the total number of monthly payments in the five years in which Caitlin plans to repay the loan, so $n = 5 \cdot 12 = 60$. Let's substitute the values of r and n into the formula.

$$P = \dfrac{0.003A}{1-\left(1+0.003\right)^{-60}}$$

Rounding the coefficient off to the nearest thousandth, this simplifies to $P = 0.018A$. The graph of $P = 0.018A$ for $5000 \leq A \leq 50{,}000$ is shown on the following page.

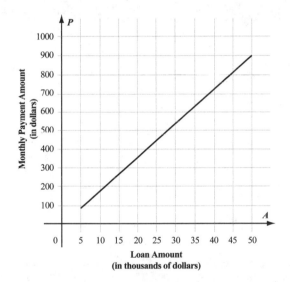

The monthly payment amount is on the vertical axis, so find the point on the line segment that aligns with 450 on the vertical axis. The A-value, or loan amount, of this point is 25, but this is in thousands of dollars. So, if Caitlin can pay $450 per month toward the loan repayment, she should take out a $25,000 loan.

To confirm our answer, let's substitute 25,000 for A in $P = 0.018A$.

$P = 0.018(25{,}000) = 450$

Yes, a $25,000 loan at the terms described results in a $450 monthly payment amount.

Lesson 2.4
Rational Expressions

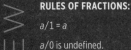
> A **rational expression** is an algebraic expression
> in the form of a ratio of two polynomials.

Keep in mind that any
polynomial is a rational
expression where the
denominator is 1.

All those familiar rules of fractions also apply to rational expressions.

For example, a ratio of the form $\dfrac{a+b}{c}$ can be rewritten as the sum

$a/c + b/c$.

In a direct variation, one variable is a constant multiple of the other variable, as in $a = kb$, where a and b are variables and k is the constant of variation. In this case, time spent setting up is equal to some constant times the square root of the number of guests, or $\sqrt{n}$.

An inverse variation between variables a and b is expressed in the form $a = k/b$. Time spent setting up is inversely proportional to the number of waiters working the event, so w must go in the denominator of the fraction set equal to time.

Only one constant of variation, c, is needed for the entire combined variation.

So, the time, t, in minutes, that it takes w waiters to set up for a dinner with n

guests is given by the equation $t = \dfrac{c\sqrt{n}}{w}$. The correct answer is (E).

EXAMPLE **8**

Rewrite the rational expression $\dfrac{6x^2+15x}{x+2}$ as the sum of a polynomial and a rational expression.

To extract a polynomial, we need part of the numerator to be an exact multiple of the denominator. Since $6x^2$ is $6x$ times x, let's use $6x(x+2)$, or $6x^2 + 12x$.

$$\dfrac{6x^2+12x+3x}{x+2}$$ 　　　Rewrite $15x$ as the sum $12x + 3x$.

$$\dfrac{6x^2+12x}{x+2} + \dfrac{3x}{x+2}$$ 　　　Separate the rational expression into a sum of two rational expressions with the same denominator.

$$\dfrac{6x(x+2)}{x+2} + \dfrac{3x}{x+2}$$ 　　　Rewrite $6x^2 + 12x$ as $6x(x+2)$.

$$6x + \dfrac{3x}{x+2}$$ 　　　Cancel out the common factor in the first rational expression.

The new expression is the sum of a polynomial ($6x$) and a rational expression $\left(\dfrac{3x}{x+2}\right)$.

Note that x cannot equal -2, because the denominator of a fraction cannot be equal to 0.

Here is how you may see rational expressions on the SAT.

Which of the following is equivalent to the expression $\dfrac{6x+5}{x-8}$?

(A) $6 + \dfrac{53}{x-8}$ 　　　　　$6 + \dfrac{53}{x-8}$

B) $6 + \dfrac{5}{x-8}$

C) $6 - \dfrac{13}{x-8}$

D) $6 - \dfrac{5}{8}$

This is the same thing
we do when rewriting
an improper numerical
fraction as a mixed
number. The fraction
25/7 is the same as
25 ÷ 7, which equals 3
with a remainder of 4 over
the divisor, or 3 4/7.

A rational expression, like a numerical fraction, represents a division of the numerator by the denominator, so one way to rewrite a rational expression is to perform the long division, writing the remainder over the divisor as the remainder rational expression.

Use long division to rewrite the rational expression $\dfrac{2x^3 + x^2 - 13x - 5}{2x^2 + 3x - 20}$.

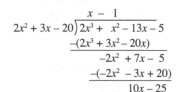

For more examples of long division of polynomials—ones without any remainders—see Example 18 in Chapter 1 and Example 4 in this chapter.

$$2x^2 + 3x - 20 \overline{\smash{\big)}\ 2x^3 + x^2 - 13x - 5}$$

$$\begin{array}{r} x \ - \ 1 \\ 2x^2 + 3x - 20 \overline{\smash{\big)}\ 2x^3 + \ x^2 - 13x - 5} \\ \underline{-(2x^3 + 3x^2 - 20x)} \\ -2x^2 \ + 7x - \ 5 \\ \underline{-(-2x^2 - 3x + 20)} \\ 10x - 25 \end{array}$$

So, we can rewrite $\dfrac{2x^3 + x^2 - 13x - 5}{2x^2 + 3x - 20}$ as the sum of the quotient and the remainder expression: $x - 1 + \dfrac{10x - 25}{2x^2 + 3x - 20}$. However, unless otherwise specified, we should always simplify a rational expression as much as possible.

$$\frac{10x - 25}{2x^2 + 3x - 20} = \frac{5(2x - 5)}{(2x - 5)(x + 4)} \qquad \text{Factor the numerator and the denominator.}$$

$$\frac{5}{x + 4} \qquad \text{Cancel out the common factor, } (2x - 5).$$

So, $\dfrac{2x^3 + x^2 - 13x - 5}{2x^2 + 3x - 20} = x - 1 + \dfrac{5}{x + 4}$, where $x \neq -4$ and $x \neq 5/2$.

The reason $x \neq 5/2$ is because $(2x - 5)$ was in the denominator before getting

canceled out. Because 5/2 was excluded from the x-value set of the original expression, it must still be excluded from the simplified expression.

OPERATIONS WITH RATIONAL EXPRESSIONS

Like rational numbers, algebraic rational expressions are closed under addition, subtraction, multiplication, and division by a non-zero rational expression. When you perform these operations on rational expressions, the result is always a rational expression.

Use the rules of operations on rational numbers (specifically fractions) to perform operations with algebraic rational expressions. For example, to add or subtract rational expressions, find the least common denominator and convert both expressions to have that denominator.

Even when the result of operations on rational expressions is a polynomial, such as $4x + 3$, this is still technically a type of rational expression ($\frac{4x+3}{1}$).

We want to rewrite the rational expression $\frac{6x+5}{x-8}$ as the sum of 6 and a leftover rational expression. To do this, we need to break the given rational expression into a sum of two rational expressions, one of which is equal to 6. The denominator is $x - 8$, so a rational expression equal to 6 and having this denominator is $\frac{6(x-8)}{x-8}$, or $\frac{6x-48}{x-8}$. We need to include $6x - 48$ in the numerator but keep the total value of $6x + 5$. In other words, we must rewrite the numerator $6x + 5$ as $6x - 48 + 53$. (Adding 53 to −48 produces 5.)

$\dfrac{6x-48+53}{x-8}$	Rewrite the numerator to include $6x - 48$.
$\dfrac{6x-48}{x-8} + \dfrac{53}{x-8}$	Rewrite the rational expression as a sum of two expressions.
$\dfrac{6(x-8)}{x-8} + \dfrac{53}{x-8}$	Factor 6 out of the numerator in the first rational expression.
$6 + \dfrac{53}{x-8}$	Cancel the common factor of $(x - 8)$.

Alternatively, we could use long division to rewrite the given rational expression.

$$
\begin{array}{r}
6 \\
x-8 \overline{)\,6x + 5} \\
-(6x - 48) \\
\hline
53
\end{array}
$$

So, $(6x + 5)$ divided by $(x - 8)$ is equal to 6 with a remainder of 53, or $6 + \dfrac{53}{x-8}$. The correct answer is (A).

Subtract: $\dfrac{x+9}{x^2-9} - \dfrac{6}{x^2-3x}$

First, let's factor each denominator, so that we can identify any common factors.

$$\frac{x+9}{(x+3)(x-3)} - \frac{6}{x(x-3)}$$

The denominators share a common factor of $(x-3)$. The least common multiple of x^2-9 and x^2-3x is $x(x+3)(x-3)$, so the least common denominator is $x(x+3)(x-3)$. To convert the first term to have that denominator, we must multiply its numerator and its denominator by x. To convert the second term, we must multiply its numerator and its denominator by $(x+3)$.

$$\frac{x(x+9)}{x(x+3)(x-3)} - \frac{6(x+3)}{x(x+3)(x-3)}$$

Convert each rational expression to have a denominator of $x(x+3)(x-3)$.

$$\frac{x^2+9x}{x(x+3)(x-3)} - \frac{6x+18}{x(x+3)(x-3)}$$

Expand both numerators.

$$\frac{x^2+9x-(6x+18)}{x(x+3)(x-3)}$$

Rewrite as a single rational expression.

$$\frac{x^2+9x-6x-18}{x(x+3)(x-3)}$$

Distribute the negative sign through to both terms in parentheses.

$$\frac{x^2+3x-18}{x(x+3)(x-3)}$$

Combine like terms.

$$\frac{(x+6)(x-3)}{x(x+3)(x-3)}$$

Factor the numerator.

$$\frac{x+6}{x(x+3)}$$

Cancel out the common factor of $(x-3)$.

The difference is $\dfrac{x+6}{x(x+3)}$, or $\dfrac{x+6}{x^2+3x}$, with $x \neq 3$, $x \neq 0$, and $x \neq -3$.

To multiply two rational expressions, multiply the numerators and multiply the denominators. To divide one rational expression by another, multiply the dividend expression by the reciprocal of the divisor expression. In either situation, remember to then simplify by cancelling out any common factors in the numerator and denominator of the product.

EXAMPLE 11

Paula, Peyton, and Phaedra pick apples at an orchard. Paula picks at a rate of b bushels per hour. Peyton picks 5 more bushels per hour than Paula. Phaedra picks twice as many bushels per hour as Paula. Yesterday all three of them picked apples. Today only Paula and Peyton are picking apples. How many times as long will it take just Paula and Peyton to pick 120 bushels as it would for Paula, Peyton, and Phaedra working together? Express your answer as a rational expression in terms of b. At what apple-picking rate of Paula's would it take twice as long for Paula and Peyton to pick 120 bushels as all three pickers working together? At what rate of Paula's would it take 1 1/2 times as long for Paula and Peyton to pick 120 bushels as all three pickers working together?

b = Paula's rate, in bushels per hour

$b + 5$ = Peyton's rate, in bushels per hour

$2b$ = Phaedra's rate, in bushels per hour

The time it takes to pick 120 bushels is the total number of bushels (120) divided by the rate, in bushels per hour. The rate for Paula and Peyton working together is $b + b + 5$, or $2b + 5$. So, the time it takes Paula and Peyton to pick 120 bushels together is $\dfrac{120}{2b+5}$.

This is a variation on the $d = rt$, or $t = d/r$, formula. In this case, the "distance" is instead total number of apples picked.

The rate for all three people working together is $b + b + 5 + 2b$, or $4b + 5$. The time it takes Paula, Peyton, and Phaedra to pick 120 bushels is $\dfrac{120}{4b+5}$.

To find how many times as long it will take just Paula and Peyton to pick 120 bushels as Paula, Peyton, and Phaedra, we must divide the time for just Paula and Peyton by the time for all three pickers.

$$\frac{120}{2b+5} \div \frac{120}{4b+5} = \frac{120}{2b+5} \times \frac{4b+5}{120}$$

Express as multiplication by the reciprocal of the divisor.

$$= \frac{120\left(4b+5\right)}{120\left(2b+5\right)}$$

Multiply the rational expressions.

$$= \frac{4b+5}{2b+5}$$

Cancel out the common factor of 120.

To find out at what apple-picking rate of Paula's it would take twice as long for just the two pickers to pick 120 bushels, as compared to all three pickers working together, we must find the b-value that makes the ratio of times equal to 2.

$\dfrac{4b+5}{2b+5} = 2$	Set the ratio of times equal to 2.
$4b + 5 = 2(2b + 5)$	Multiply both sides by $(2b + 5)$.
$4b + 5 = 4b + 10$	Distribute.
$4b - 4b = 10 - 5$	Move b terms to one side and numbers to the other.
$0 = 5$	Simplify.

This equation is not true, so there is no rate b such that it will take just Paula and Peyton twice as long as all three of them.

Find at what rate of Paula's (what value of b) it will take 1 1/2 times as long for Paula and Peyton to pick 120 bushels as it would take all three pickers working together.

$\dfrac{4b+5}{2b+5} = 3/2$	Rewrite 1 1/2 as 3/2.
$2(4b + 5) = 3(2b + 5)$	Cross-multiply.
$8b + 10 = 6b + 15$	Distribute.
$8b - 6b = 15 - 10$	Move b terms to one side and numbers to the other.
$2b = 5$	Subtract.
$b = 5/2$, or 2.5	Divide both sides by 2.

If Paula picks apples at a rate of 2.5 bushels per hour, then it will take Paula and Peyton 1 1/2 times as long to pick 120 bushels as Paula, Peyton, and Phaedra all working together.

CHAPTER 2 PRACTICE QUESTIONS

Directions: Complete the following open-ended problems as specified by each question stem. For extra practice after answering each question, try using an alternative method to solve the problem or check your work.

1. The variable b varies inversely as a, and $b = 13.5$ when $a = 4.5$. Find the constant of variation, write an equation for the relationship, and find b when a is 0.5.

   $C = 60.75$ $b = 121.5$

2. Write an equation for the volume of a rectangular prism with a length of l inches, a width of w inches, and a height of 8 inches. Identify the type of variation and the constant of variation, then find the volume of the prism if the length of the base is 4 inches and the width of the base is 2 inches. $V = 8lw$ direct var.

   $V = 64 \text{ in}^3$

3. What are the roots of the polynomial $2x^3 + 26ix^2 - 44x$?

   $0 , -2i , -11i$

4. Simplify $\dfrac{x+1}{x^2 + 2x - 3} \cdot \dfrac{x^2 + x - 6}{x^2 - 2x - 3}$ $\dfrac{x-2}{(x-1)(x-3)}$

5. How many roots does $x^3 - 5x^2 + 144x - 720$ have? What are the roots?

   $3 : 5 , 12i , -12i$

6. Solve the equation $x^5 - x^4 = 2x^3 + 4x^2 + 24x$.

7. Simplify the expression

   $\dfrac{3}{x^2 + x} + \dfrac{x+4}{x^2 + 2x + 1}$.

8. The monthly payment, m, on a mortgage varies directly with the amount borrowed, A. If the monthly payment on a 30-year mortgage is $7.82 for every $2,000 borrowed, find a formula that relates the monthly payment to the amount borrowed for a mortgage with these terms. Then find the monthly payment when the amount borrowed is $120,000.

9. A box has a volume given by $2x^3 - 3x^2 - 39x + 20$. If the height of the box is given by $2x - 1$, what is the area of the base? If the length is given by $x + 2$, find a sum of a polynomial and a rational expression that would be equivalent to the width.

4.

5. 5 | 1   -5   144   -720
          5   0   720
      1   0   144   0

   $x^2 + 144$

6. $x^5 - x^4 - 2x^3 - 4x^2 - 24x = 0$

   $x(x^4 - x^3 - 2x^2 - 4x - 24) = 0$

   -3 | 1   -1   -2   -4   -24

---

1. $b = \dfrac{C}{a}$

   $C = 60.75$

   $b = 121.5$

2. $V = 8lw$

   $V = 64$

3. $2x(x^2 + 13ix - 22)$

   $2x(x + 2i)(x + 11i)$

   $(x+1)(x+3)(x-2)$

   $\dfrac{(x+3)(x-1)(x-3)(x+1)}{}$

   $\dfrac{x-2}{(x-1)(x-3)}$

SOLUTIONS TO CHAPTER 2 PRACTICE QUESTIONS

1. **constant of variation: 60.75; equation: ab = 60.75; b = 121.5**
   Start with the expression for inverse variation: $ab = k$ and make the appropriate substitutions to evaluate the constant of variation.
   $$ab = k$$
   $$(4.5)(13.5) = k$$
   $$k = 60.75$$

   Thus, an equation for the relationship is ab = 60.75. Since we now must solve for b, it can be helpful to rewrite the equation as b = 60.75/a. Plug in a = 0.5 and evaluate for b.
   $$b = 60.75/a$$
   $$b = 60.75/(0.5)$$
   $$b = 121.5$$

2. **64**
   The volume of a rectangular prism is $V = lwh$.
   $$V = lwh$$
   $$V = lw(8), \text{ or } 8lw$$

   Thus, volume varies jointly with the length, l, and width, w. The constant of variation is 8.

   Next, use the provided dimensions to calculate the volume of the prism.
   $$V = 8lw = 8(4)(2) = 64$$

   The volume is 64 cubic inches.

3. **0, −11i, and −2i**
   Since this polynomial is a cubic function (degree 3), it will have three total roots. First, factor out the common term of $2x$; this will give $(2x)(x^2 + 13ix − 22)$. Since there is an i term in the coefficient, rewrite −22 in terms of i; this gives the form $(2x)[x^2 + 13ix + (−22)]$, or $(2x)[x^2 + 13ix + (22i^2)]$. This can be factored into the form $(2x)(x + 11i)(x + 2i)$, giving the three roots of 0, −11i, and −2i.

4. $\dfrac{x - 2}{x^2 - 4x + 3}$

   Begin by factoring the numerators and denominators to cancel out common factors.

$\dfrac{x + 1}{x^2 + 2x - 3} \cdot \dfrac{x^2 + x - 6}{x^2 - 2x - 3}$	Given.
$\dfrac{x + 1}{(x + 3)(x - 1)} \cdot \dfrac{(x + 3)(x - 2)}{(x - 3)(x + 1)}$	Factor numerator and denominator.
$\dfrac{x - 2}{(x - 1)(x - 3)}$	Cancel out common factors.
$\dfrac{x - 2}{x^2 - 4x + 3}$	Use FOIL in the denominator.

5. **3 roots; 5, 12*i*, and −12*i***

   Since the expression has a degree of 3, the expression has 3 roots. At first, this expression does not look easy to factor; however, by grouping the first two terms and the second two terms, they can be factored.

   $x^3 - 5x^2 + 144x - 720 = 0$
   $(x^3 - 5x^2) + (144x - 720) = 0$
   $x^2(x - 5) + 144(x - 5) = 0$
   $(x^2 + 144)(x - 5) = 0$
   $x^2 + 144 = 0 \qquad x - 5 = 0$
   $x^2 = -144 \qquad\quad x = 5$
   $x = \pm 12i$

   So, the solutions are 5, 12*i*, and -12*i*.

6. **−2, 0, 3, 2*i*, and −2*i***

   This can be rewritten as a system of equations by setting each side of the equation equal to *y*, as shown here:

   $y = x^5 - x^4$
   $y = 2x^3 + 4x^2 + 24x$

   Any *x*-value that produces the same *y*-value in each equation would be a solution; therefore, graph the equations. The graph would look something like the following:

   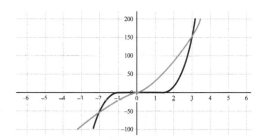

   There appear to be three solutions shown, at *x*-values of −2, 0, and 3. If tested in the original equation, these would indeed work; an *x*-value of −2 would give −48 = −48, a *y*-value of 0 would give 0 for both, and an *x*-value of 3 would give 162 for both. Therefore, those are three solutions.

   However, the highest exponent shown is 5, so there should be a total of 5 solutions. Start by moving everything over to one side of the equation: $x^5 - x^4 - 2x^3 - 4x^2 - 24x = 0$

   The three roots found go with factors of *x*, (*x* + 2), and (*x* − 3); multiply those together.
   $x(x + 2)(x - 3) = (x^2 + 2x)(x- 3)= x^3 - 3x^2 + 2x^2 - 6x = x^3 - x^2 - 6x$

Divide this into the original equation to find what is left over:

$$x^3 - x^2 - 6x \overline{\smash{\big)}\ x^5 - x^4 - 2x^3 - 4x^2 - 24x} \quad \overset{\displaystyle x^2 \quad + \quad 4}{}$$

$$\underline{-\,(x^5 - x^4 - 6x^3)}$$
$$4x^3 - 4x^2 - 24x$$
$$\underline{-\,(4x^3 - 4x^2 - 24x)}$$
$$0$$

The final factor is $(x^2 + 4)$, which has roots $2i$ and $-2i$. Therefore, the five solutions to the system of equations are -2, 0, 3, $2i$, and $-2i$.

7. $\dfrac{x^2 + 7x + 3}{x(x + 1)^2}$

First, factor the denominators to determine the Lowest Common Denominator so it will be

possible to add the numerators. The first denominator, $x^2 + x$, factors to $x(x + 1)$, and the second

denominator, $x^2 + 2x + 1$, factors to $(x + 1)^2$. Therefore, the least common denominator possible

here is $x(x + 1)^2$. Multiply the first fraction by $\dfrac{x + 1}{x + 1}$ and the second fraction by $\dfrac{x}{x}$ so both

rational expressions will have the same denominator. This will give $\dfrac{3(x + 1)}{x(x + 1)^2} + \dfrac{x(x + 4)}{x(x + 1)^2}$.

Since the denominators are now the same, the numerators can be added above the common

denominator; this gives $\dfrac{3(x + 1) + x(x + 4)}{x(x + 1)^2}$. Given there are products to be found in the

numerator, distribute to simplify; the result is $\dfrac{3x + 3 + x^2 + 4x}{x(x + 1)^2}$. Finally, combine like terms to

get the final answer of $\dfrac{x^2 + 7x + 3}{x(x + 1)^2}$. Since this final numerator does not factor there is no more

simplification that can be done here, so the final answer is $\dfrac{x^2 + 7x + 3}{x(x + 1)^2}$.

8. **\$469.20**
   Because m varies directly with A, we know that $m = kA$. Use the given values to determine the constant of variation.
   $m = kA$
   $7.82 = k(2000)$
   $k = 0.00391$

   Thus, the formula is $m = 0.00391A$. Next, use this to determine m when $A = 120{,}000$.
   $m = 0.00391A$
   $m = 0.00391(120{,}000)$
   $m = 469.2$
   So, when the amount borrowed is \$120,000, the monthly payment would be \$469.20.

9. $x - 3 - \dfrac{14}{x+2}$

Since volume of a box is given by lwh and the area of the base of a box is given by lw, this can

be set up as $\dfrac{2x^3 - 3x^2 - 39x + 20}{2x - 1}$. To find the area of the base, complete the division:

$$
\begin{array}{r}
x^2 - x - 20 \\
2x - 1 \overline{\smash{)}\, 2x^3 - 3x^2 - 39x + 20} \\
\underline{-\,(2x^3 - x^2)} \\
-2x^2 - 39x \\
\underline{-\,(-2x^2 + x)} \\
40x + 20 \\
\underline{-(-40x + 20)} \\
0
\end{array}
$$

This gives the area of the base as $x^2 - x - 20$. However, as the question also wants the width of

the box, this quadratic needs to be further broken down. To find the width, you must divide the

quadratic by the length, which is $x + 2$. Write this as a rational expression: $\dfrac{x^2 - x - 20}{x + 2}$. To extract a

polynomial term, this needs to be rewritten in a form to allow us to extract something; for example,

rewrite the numerator as $x^2 + 2x - 3x - 20$. This would allow the expression to be broken down

as $\dfrac{x^2 + 2x}{x + 2} - \dfrac{3x + 20}{x + 2}$. The first term would be factored as $\dfrac{x(x+2)}{x+2}$, which is just x. The second

piece can further be broken down by rewriting $\dfrac{3x + 20}{x + 2}$ as $\dfrac{3x + 6 + 14}{x + 2}$; this would break into

$\dfrac{3x + 6}{x + 2} + \dfrac{14}{x + 2}$, or $\dfrac{3(x+2)}{x+2} + \dfrac{14}{x+2}$.

The $x + 2$ terms cancel in the first piece here, leaving just $3 + \dfrac{14}{x+2}$. Putting everything together,

this gives the expression $x - (3 + \dfrac{14}{x+2})$, which simplifies as $x - 3 - \dfrac{14}{x+2}$.

REFLECT

**Congratulations on completing Chapter 2!
Here's what we just covered.
Rate your confidence in your ability to:**

- Determine how many roots a polynomial has, based on the degree of the polynomial

① ② ③ ④ ⑤

- Factor polynomials with imaginary number coefficients

① ② ③ ④ ⑤

- Create and solve systems of equations to solve single-variable polynomial equations

① ② ③ ④ ⑤

- Identify and apply direct, inverse, joint, and combined variations

① ② ③ ④ ⑤

- Perform arithmetic operations with rational expressions

① ② ③ ④ ⑤

- Rewrite rational expressions as the sum of a polynomial and a remainder expression

① ② ③ ④ ⑤

If you rated any of these topics lower than you'd like, consider reviewing the corresponding lesson before moving on, especially if you found yourself unable to correctly answer one of the related end-of-chapter questions.

Access your online student tools for a handy, printable list of Key Points for this chapter. These can be helpful for retaining what you've learned as you continue to explore these topics.

Chapter 3
Radical and Rational Equations and Inequalities

GOALS
By the end of this chapter, you will be able to

- Solve single-variable rational and radical equations and inequalities

- Write and solve rational and radical equations, inequalities, and functions to describe real-life situations and solve problems

- Graph rational and radical functions

- Use systems of equations to solve single-variable rational and radical equations

Lesson 3.1
Rational Equations in One Variable

A **rational equation** is an equation that contains one or more rational expressions. These equations can be used to solve problems that involve ratio comparisons or division, such as the calculation of an average.

EXAMPLE 1

Will scored a 90 on the math test he took today, which brought his average (arithmetic mean) test grade for the semester up from 78 to 80. How many math tests has he taken this semester, including today's?

Calculating a mean involves division, so we'll write a rational equation to represent this situation. Will's mean test score for the semester is the total value of all of his test scores divided by the number of tests he has taken.

$$\text{mean score} = \frac{\text{total value of scores}}{\text{number of tests}}$$

Let n = the number of math tests Will took **before** today's test. His average for those first n tests was 78, so $78 = \dfrac{\text{total of first } n \text{ scores}}{n}$. If you rewrite this equation (multiply both sides by n), you can see that the total of Will's scores on the first n tests was $78n$.

Let's write an equation for the average of all of Will's test scores, including today's. He took a total of $n + 1$ tests. The total of all of his test scores is the sum of the total of his first n test scores ($78n$) and his test score from today (90). His new average test score is 80.

$$80 = \frac{78n + 90}{n + 1}$$

Solve the equation for n.

$80(n + 1) = 78n + 90$	Multiply both sides by $(n + 1)$.
$80n + 80 = 78n + 90$	Distribute through both terms in parentheses.
$2n = 10$	Move all n terms to one side and all constants to the other.
$n = 5$	Divide both sides by 2.

This isn't yet our answer, though! Remember, n is the number of tests Will took **before** today's test, but the question asks for the total number of tests **including** today's. Will has taken a total of $5 + 1 = 6$ math tests this semester.

If a rational equation shows one rational expression set equal to another, you can cross-multiply and solve the resulting polynomial equation.

 Here's your chance to use all the polynomial-solving techniques you learned in Chapter 1!

EXAMPLE

Solve for any real solutions to the following equation.

$$\frac{2x+1}{x^2 - 8x + 50} = \frac{1}{x^2 - 4}$$

Cross-multiply and move all terms to one side.

$(2x + 1)(x^2 - 4) = 1(x^2 - 8x + 50)$
$2x^3 + x^2 - 8x - 4 = x^2 - 8x + 50$
$2x^3 - 54 = 0$

The shortcut here is to recognize that this is the simple cubic equation $2x^3 = 54$, or $x^3 = 27$, which has the solution $x = 3$. If we instead factor $2x^3 - 54$, we can solve for all three solutions.

$2(x^3 - 27) = 0$	Factor out the common factor, 2.
$2(x - 3)(x^2 + 3x + 9) = 0$	Factor the difference of cubes.

The solution to $x - 3 = 0$ is $x = 3$. Solving $x^2 + 3x + 9 = 0$ requires the quadratic formula.

$$x = \frac{-3 \pm \sqrt{3^2 - 4(1)(9)}}{2(1)} = \frac{-3 \pm \sqrt{-27}}{2} = \frac{-3 \pm 3i\sqrt{3}}{2}$$

These two solutions include imaginary numbers, so the only real solution is 3. Test it in the original equation.

$$\frac{2(3)+1}{3^2 - 8(3) + 50} \overset{?}{=} \frac{1}{3^2 - 4}$$

$$7/35 \overset{?}{=} 1/5$$

$$1/5 = 1/5$$

The equation is true, so $x = 3$ is the one real solution to $\dfrac{2x+1}{x^2 - 8x + 50} = \dfrac{1}{x^2 - 4}$.

Testing your answer in the original equation is an especially good idea when it comes to rational equations, because of the risk of false solutions.

> An x-value that makes the denominator of a rational expression equal to zero makes the entire rational expression undefined and therefore cannot be a solution to the given rational equation. However, such x-values may be solutions to the polynomial equations produced during the solution process, through cross-multiplication or other methods. These false solutions are called **extraneous solutions**.

You'll specifically want to keep this in mind when dealing with multiple-choice problems like the ones found on the SAT, as extraneous solutions are often offered as trap answers.

Sometimes a rational equation will have multiple potential solutions, including both extraneous solutions and true solutions. You will need to determine which are which by testing each potential solution in the original equation.

Solve: $\dfrac{x+2}{x} - \dfrac{x+9}{x+6} = \dfrac{2-x}{x}$

The least common denominator for all three rational expressions is $x(x + 6)$. Convert all three expressions to have this denominator.

$$\dfrac{(x+2)(x+6)}{x(x+6)} - \dfrac{x(x+9)}{x(x+6)} = \dfrac{(2-x)(x+6)}{x(x+6)}$$

$$\dfrac{x^2+8x+12}{x(x+6)} - \dfrac{x^2+9x}{x(x+6)} = \dfrac{-x^2-4x+12}{x(x+6)}$$

Expand the numerators.

In Lesson 2.4, we practiced operations on rational expressions. Here, we're using those skills to combine rational expressions in order to solve a rational equation.

$$\dfrac{x^2+8x+12-(x^2+9x)}{x(x+6)} = \dfrac{-x^2-4x+12}{x(x+6)}$$

Combine the two rational expressions on the left side.

$$\dfrac{x^2+8x+12-x^2-9x}{x(x+6)} = \dfrac{-x^2-4x+12}{x(x+6)}$$

Distribute the negative sign to both terms in parentheses.

$$\dfrac{-x+12}{x(x+6)} = \dfrac{-x^2-4x+12}{x(x+6)}$$

Combine like terms in the numerator.

Here is how you may see rational equations on the SAT.

$$\dfrac{1}{r-3} + \dfrac{1}{r+5} = \dfrac{8}{(r-3)(r+5)}$$

Given the equation above, what is the value of r?

A) 3

B) 6

C) 7

D) There is no such value of r.

The two rational expressions have the same denominator, so their numerators must be equal.

Setting the numerators equal is the same as multiplying both sides by $x(x + 6)$. You could also have multiplied every term on both sides by $x(x + 6)$ at the beginning of the solution process.	$-x + 12 = -x^2 - 4x + 12$	Set the numerators equal to one another.
	$x^2 + 3x = 0$	Move all terms to the left side and combine like terms.
	$x(x + 3) = 0$	Factor the quadratic.
	$x = 0$ or $x = -3$	Set each factor equal to 0 and solve for x.

Test each of the potential solutions in the original equation.

When $x = 0$: $\dfrac{0+2}{0} - \dfrac{0+9}{0+6} \stackrel{?}{=} \dfrac{2-0}{0}$

Two of the rational expressions have 0 in the denominator, so these are undefined. The x-value of 0 is an extraneous solution. It is **not** a solution to the rational equation.

When $x = -3$: $\dfrac{-3+2}{-3} - \dfrac{-3+9}{-3+6} \stackrel{?}{=} \dfrac{2-(-3)}{-3}$

$$1/3 - 6/3 \stackrel{?}{=} -5/3$$

$$-5/3 = -5/3$$

This equation is true, so -3 is a solution for x. The only solution to $\dfrac{x+2}{x} - \dfrac{x+9}{x+6} = \dfrac{2-x}{x}$ is $x = -3$.

3

EXAMPLE 4

Solve the following equation for x.

$$\frac{x-2}{2x+4} = \frac{2-x}{x^2+2x}$$

Let's cross-multiply then solve for possible solutions for x.

$(x - 2)(x^2 + 2x) = (2 - x)(2x + 4)$

$x^3 + 2x^2 - 2x^2 - 4x = 4x + 8 - 2x^2 - 4x$	Use FOIL to expand.
$x^3 - 4x = -2x^2 + 8$	Combine like terms on each side.
$x^3 + 2x^2 - 4x - 8 = 0$	Move all terms to one side of the equation.

We need to factor this cubic to solve for all possible values of x. According to the Rational Root Theorem, any rational roots of this cubic will be in the set $\pm\{1, 2, 4, 8\}$. Let's use the Remainder Theorem to test some values in the polynomial $x^3 + 2x^2 - 4x - 8$.

When $x = 1$, $1^3 + 2(1^2) - 4(1) - 8 = -9$. This is not equal to 0, so 1 is not a root of the polynomial.

When $x = 2$, $2^3 + 2(2^2) - 4(2) - 8 = 0$. This means that 2 is a root, so $(x - 2)$ is a factor of the cubic. We can use long division to determine the other factor.

The Rational Root Theorem and the Remainder Theorem are what we used to find the roots of a cubic polynomial in Chapter 1, Example 18. Here, we are using the same methods to solve for zeros of a polynomial equation, which in turn are potential solutions to the given rational equation.

$$\begin{array}{r} x^2 + 4x + 4 \\ x-2 \overline{\smash{\big)}\ x^3 + 2x^2 - 4x - 8} \\ \underline{-(x^3 - 2x^2)} \\ 4x^2 - 4x \\ \underline{-(4x^2 - 8x)} \\ 4x - 8 \\ \underline{-(4x - 8)} \\ 0 \end{array}$$

We can start by combining the two rational expressions on the left side of the equation. To do so, we must convert them to have their least common denominator, $(r - 3)(r + 5)$, then add them together.

$$\frac{r+5}{(r-3)(r+5)} + \frac{r-3}{(r-3)(r+5)} = \frac{8}{(r-3)(r+5)}$$

$$\frac{r+5+r-3}{(r-3)(r+5)} = \frac{8}{(r-3)(r+5)}$$

We don't need to cross-multiply, because the two rational expressions that are equal to one another have the same denominator. This means that their numerators are equal.

$$r + 5 + r - 3 = 8$$
$$2r + 2 = 8$$
$$2r = 6$$
$$r = 3$$

It looks like we found the solution!...Or is it an extraneous solution masquerading as a solution? Before you choose (A), test $r = 3$ in the original equation.

$$\frac{1}{3-3} + \frac{1}{3+5} \stackrel{?}{=} \frac{8}{(3-3)(3+5)}$$

$$1/0 + 1/8 \stackrel{?}{=} 8/0$$

Two of the fractions have denominators of 0, which means that they have undefined values. So, 3 is **not** a solution to the original rational equation. It is an extraneous solution. There are no other possible solutions to this equation, so the correct answer is (D).

We can rewrite our equation $x^3 + 2x^2 - 4x - 8 = 0$ using this factorization.

$$(x - 2)(x^2 + 4x + 4) = 0$$
$$(x - 2)(x + 2)(x + 2) = 0 \qquad \text{Factor the quadratic.}$$

The potential solutions are $x = 2$ and $x = -2$. Test each solution in the original rational equation, $\dfrac{x-2}{2x+4} = \dfrac{2-x}{x^2+2x}$.

When $x = 2$:
$$\frac{2-2}{2(2)+4} \overset{?}{=} \frac{2-2}{2^2+2(2)}$$
$$0/8 \overset{?}{=} 0/8$$
$$0 = 0$$

This numerical equation is true, so 2 is a solution for x.

When $x = -2$:
$$\frac{-2-2}{2(-2)+4} \overset{?}{=} \frac{2-(-2)}{(-2)^2+2(-2)}$$
$$-4/0 \overset{?}{=} 4/0$$

The expressions $-4/0$ and $4/0$ are both undefined, so -2 is an extraneous solution.

The only solution to the given rational equation is $x = 2$.

An alternative means of factoring the cubic could have involved grouping out the factors in $x^3 + 2x^2 - 4x - 8 = 0$, so as to get $x^2(x + 2) - 4(x + 2) = 0$, which equals $(x^2 - 4)$ $(x + 2) = 0$. Higher-degree expressions can often be manipulated in this fashion, and if you can spot these commonalities, it may save you some time.

Factoring at the beginning can often help with the solution process. If we had first factored the denominators, we would have gotten the following.

$$\frac{x-2}{2(x+2)} = \frac{2-x}{x(x+2)}$$

There is a factor of $(x + 2)$ in both denominators. We can multiply both sides of the equation by $(x + 2)$ and cancel out the factor $(x + 2)$ in each denominator, as long as we note that $x + 2 \neq 0$, so $x \neq -2$. Here is the resulting equation.

$$\frac{x-2}{2} = \frac{2-x}{x} \quad \text{where } x \neq -2$$

When we cross-multiply this equation, we get $x^2 - 2x = 4 - 2x$, which simplifies to $x^2 = 4$. The two solutions to $x^2 = 4$ (which can also be written as $x^2 - 4 = 0$) are $x = 2$ and $x = -2$. But, we established earlier that x cannot equal -2. That means that the only solution is $x = 2$. When we test 2 in the original equation, it produces a correct numerical equation, as shown above.

This method is simpler than the first approach we used, involving only a quadratic equation instead of a cubic equation, but it requires making sure the denominator of each rational expression never equals 0.

Because x is another factor in one of the denominators, it is also true that x cannot equal 0. However, 0 was not a solution to the polynomial equation we found through cross-multiplying, so it did not come up as an extraneous solution.

Lesson 3.2
Rational Inequalities in One Variable

If you multiply or divide both sides of an inequality by a negative number, you must switch the direction of the inequality sign.

If you rewrite an inequality statement with the sides reversed, you must switch the direction of the inequality sign; if $a > b$, then $b < a$.

When graphing an inequality solution set on a number line, use a solid circle to indicate that an endpoint is included in the solution set (corresponding to the signs $\geq$ and $\leq$) and an open circle to indicate that an endpoint is not included in the solution set (corresponding to the signs $>$ and $<$).

A **rational inequality** is an inequality that contains one or more rational expressions. Be sure to follow the same rules for inequality signs that you would with numerical inequalities, when dealing with algebraic expressions.

EXAMPLE 5

Kristi is going for a one-hour run. She covered the first mile in 8 minutes. How many more miles must she run if she wants to average less than 11 minutes per mile for the entire run?

The key words "less than" mean that we should set up an inequality to solve this problem. Kristi wants her average to be less than 11 minutes per mile, so the inequality will end in < 11. Her average minutes per mile is her total number of minutes spent running divided by her total number of miles run. Her total number of minutes is 60, because she is going for a one-hour run and there are 60 minutes in one hour. The unknown, which we can call m, is the number of miles left in her run, but she has already run one mile. So, her total number of miles in the run is $m + 1$. Let's write our inequality.

$$\frac{60}{m+1} < 11$$

Notice that the fact that her first mile took 8 minutes is extraneous information—not necessary for writing or solving this particular inequality. Some tests provide extraneous information to throw you off.

Solve the inequality for m.

$$60 < 11(m + 1)$$
$$60 < 11m + 11$$
$$49 < 11m$$
$$49/11 < m$$
$$4\,5/11 < m$$

The solution for m is $m > 4\,5/11$. Kristi must run more than $4\,5/11$ more miles during this run if she wants to average less than 11 minutes per mile for the entire run.

The rational inequality in Example 5 was simple to solve algebraically, but most rational inequalities cannot be solved solely using the processes we use for rational equations.

> To solve a rational inequality, rewrite as a rational expression set across an inequality symbol from 0, find the zeros and points of discontinuity of the rational expression, and find which intervals between those points make the inequality true.

EXAMPLE 6

Solve the following inequality.

$$\frac{x}{x+5} \geq \frac{2}{x-1}$$

One way to approach this is to temporarily solve as an equation, cross-multiplying to get $x^2 - x = 2x + 10$, moving all terms to one side to get $x^2 - 3x - 10 = 0$, and solving for the zeros of the polynomial. But we would need to also remember that $x \neq -5$ and $x \neq 1$ from the original equation and use these values of x along with the zeros as points to divide the number line into intervals. Another approach is to subtract $\dfrac{2}{x-1}$ from both sides.

$$\frac{x}{x+5} - \frac{2}{x-1} \geq 0$$

$$\frac{x(x-1)}{(x-1)(x+5)} - \frac{2(x+5)}{(x-1)(x+5)} \geq 0$$

Convert both expressions to have their least common denominator.

$$\frac{x^2 - x - (2x + 10)}{(x-1)(x+5)} \geq 0$$

Subtract to combine the rational expressions.

$$\frac{x^2 - 3x - 10}{(x-1)(x+5)} \geq 0$$

Distribute the subtraction and combine like terms.

$$\frac{(x+2)(x-5)}{(x-1)(x+5)} \geq 0$$

Factor the quadratic in the numerator.

The rational expression $\frac{(x+2)(x-5)}{(x-1)(x+5)}$ will equal 0 if either factor in the numerator

equals 0. When $x + 2 = 0$, $x = -2$, and when $x - 5 = 0$, $x = 5$. If either factor in the

denominator equals 0, then the entire denominator equals 0 and the rational

expression is undefined. So, $x \neq 1$ and $x \neq -5$. The two values that make the

expression equal 0 and the two values that make it undefined are the four x-values

that define the intervals we will look at. Let's plot these four points on a number

line. The points at −5 and 1 must be open circles, because these values of x are **not**

included in the solution set. The points at −2 and 5 are closed circles, because these

endpoints are included in the solution set by the $\geq$ sign (as opposed to the > sign).

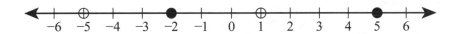

Let's test a value from each interval in the original inequality.

When $x = -6$: $\dfrac{-6}{-6+5} \geq \dfrac{2}{-6-1} \rightarrow 6 \geq -2/7$ True

When $x = -3$: $\dfrac{-3}{-3+5} \geq \dfrac{2}{-3-1} \rightarrow -3/2 \geq -1/2$ False

When $x = 0$: $\dfrac{0}{0+5} \geq \dfrac{2}{0-1} \rightarrow 0 \geq -2$ True

When $x = 3$: $\dfrac{3}{3+5} \geq \dfrac{2}{3-1} \rightarrow 3/8 \geq 1$ False

When $x = 10$: $\dfrac{10}{10+5} \geq \dfrac{2}{10-1} \rightarrow 2/3 \geq 2/9$ True

You can choose any value you like from each of the intervals, so it is helpful to choose values that will provide simpler calculations. For example, testing $x = 10$ produces an easier comparison than testing $x = 8$, which would give us the inequality $8/13 \geq 2/7$.

The inequality is true for *x*-values less than −5, greater or equal to −2 but less than 1, and greater than 5. Now we can graph the full solution set, as shown on the number line below.

We can also write the solution set as "$x < -5$, $-2 \leq x < 1$, or $x \geq 5$."

Lesson 3.3
Rational Functions

REVIEW

The **domain** of a function is the set of *x*-values, or input values, for which the function exists.

An **even function** is a function $f(x)$ for which $f(-x) = f(x)$. The graph of an even function is symmetrical with respect to the *y*-axis.

An **inverse variation** is a variation equation in which one variable is inversely proportional to the other variable. If *a* is inversely proportional to *b*, then $a = k/b$, where *k* is some constant.

An **odd function** is a function $f(x)$ for which $f(-x) = -f(x)$. The graph of an odd function is symmetrical with respect to the origin.

The **range** of a function is the set of output values (generally *y*-values or $f(x)$-values) corresponding to the *x*-values in the domain of the function. The *x*- and *y*-intercepts are points of intersection within the *x*- and *y*-axis (respectively).

A **rational function** is a function that is defined as a rational expression, a ratio of one polynomial to another.

One very simple type of rational function is an inverse variation relationship.

In the function *a* = 10/*b*, what happens to *a* as *b* approaches positive and negative infinity? Graph this function.

As the value of *b* increases, 10/*b* becomes a smaller and smaller positive fraction. As *b* approaches infinity (∞), *a* approaches 0.

If *b* were a negative number decreasing to negative infinity (−∞), then *a* would be a negative number, but it would still be getting closer and closer to 0. (For example, when *b* = −1000, *a* = −1/100.)

Use the equation to find more coordinate pairs on the graph.

When *b* = 1, *a* = 10.   When *b* = −1, *a* = −10.
When *b* = 5, *a* = 2.   When *b* = −5, *a* = −2.
When *b* = 10, *a* = 1.   When *b* = −10, *a* = −1.

So, the graph of *a* = 10/*b* passes through (1, 10), (5, 2), (10, 1), (−1, −10), (−5, −2), and (−10, −1).

Let's also look at what happens when the value of *b* is close to 0. When *b* = 0, the fraction 10/*b* is undefined, so *a* does not exist when *b* = 0. As *b* approaches 0 as a positive number, the value of 10/*b* becomes very large. As *b* approaches 0 as a negative number, the value of 10/*b* is negative but with a large absolute value. So, as *b* approaches 0 from the right on the graph, *a* approaches infinity, and as *b* approaches 0 from the left, *a* approaches negative infinity.

The graph of *a* = 10/*b* is shown below.

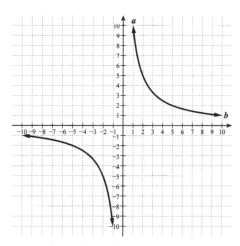

Notice that this graph is symmetrical with respect to the origin. This is because the function *a* = 10/*b* is an odd function. If you substitute −*b* for *b*, the result is −*a*.

There is a vertical asymptote at $b = 0$, or the a-axis. The function $a = 10/b$ approaches $-\infty$ on the left side and ∞ on the right side as b nears 0. The function will never intersect the a-axis. It just continually gets closer and closer to it.

There is a horizontal asymptote at $a = 0$, or the b-axis. Both ends of the function $a = 10/b$ approach this line, as b approaches $-\infty$ and ∞. The function value gets closer and closer to 0 but never reaches it.

Even if the equation in Example 7 had a more complicated denominator, such as $a = \dfrac{10}{b-5}$, both ends of the function a would still approach a horizontal asymptote of $a = 0$ as b approaches positive and negative infinity. A difference of 5 is basically meaningless when b has a huge absolute value.

Let's look at some more complicated rational functions.

Here is how you may see rational functions on the ACT.

In the equation $a = 5/b$, b is a positive real number. As the value of b is increased so it becomes closer and closer to infinity, what happens to the value of a?

    A. It remains constant.

    B. It gets closer and closer to zero.

    C. It gets closer and closer to one.

    D. It gets closer and closer to five.

    E. It gets closer and closer to infinity.

Create tables of values and graph each of the functions below.

$$f(x) = \frac{6}{x-3}$$

$$g(x) = \frac{8x}{x^2+3x-18}$$

For each rational function, we'll need more points to get a sense of the graph's behavior than we would for a linear or quadratic function.

x	$f(x) = \dfrac{6}{x-3}$	f(x)
-3	$f(-3) = \dfrac{6}{-3-3}$	-1
0	$f(0) = \dfrac{6}{0-3}$	-2
1	$f(1) = \dfrac{6}{1-3}$	-3
2	$f(2) = \dfrac{6}{2-3}$	-6
3	$f(3) = \dfrac{6}{3-3}$	undefined
4	$f(4) = \dfrac{6}{4-3}$	6
5	$f(5) = \dfrac{6}{5-3}$	3
9	$f(9) = \dfrac{6}{9-3}$	1

x	$g(x) = \dfrac{8x}{(x-3)(x+6)}$	g(x)
-9	$g(-9) = \dfrac{8(-9)}{(-9-3)(-9+6)}$	-2
-7	$g(-7) = \dfrac{8(-7)}{(-7-3)(-7+6)}$	-5.6
-6	$g(-6) = \dfrac{8(-6)}{(-6-3)(-6+6)}$	undefined
-5	$g(-5) = \dfrac{8(-5)}{(-5-3)(-5+6)}$	5
0	$g(0) = \dfrac{8(0)}{(0-3)(0+6)}$	0
2	$g(2) = \dfrac{8(2)}{(2-3)(2+6)}$	-2
3	$g(3) = \dfrac{8(3)}{(3-3)(3+6)}$	undefined
4	$g(4) = \dfrac{8(4)}{(4-3)(4+6)}$	3.2
6	$g(6) = \dfrac{8(6)}{(6-3)(6+6)}$	$1.\overline{3}$

It's clear from these tables that something funny happens to each function around the point(s) where it is undefined. The graph drops as it approaches one of those points from the left (increasing *x*-values), but on the right side it is suddenly much higher and dropping again. These places involve vertical asymptotes, such as the one in Example 7. Let's look at the complete graph of each of these functions, as shown on a graphing calculator, passing through the points we found.

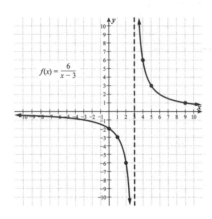

$$f(x) = \frac{6}{x - 3}$$

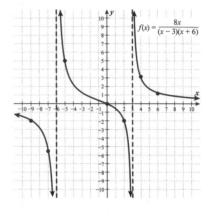

$$f(x) = \frac{8x}{(x - 3)(x + 6)}$$

To see what happens, we can plug in increasing positive values of *b* and solve for *a*.

When *b* = 1, *a* = 5.
When *b* = 5, *a* = 1.
When *b* = 50, *a* = 1/10.
When *b* = 500, *a* = 1/100.
When *b* = 5000, *a* = 1/1000.

Even though *a* reaches values of 5 and 1, which are listed in the answer choices, the question asks only for what happens to *a* as *b* approaches infinity.

As the value of *b* increases, the value of *a* decreases. It never becomes negative, though. It just becomes a smaller and smaller fraction. In other words, as *b* approaches infinity (∞), *a* approaches 0. The correct answer is (B).

The dashed lines indicate where there are vertical asymptotes. The graph approaches these lines but never actually intersects them. These are points of discontinuity for the graphs. Notice that they are exactly at the x-values where $f(x)$ and $g(x)$ are undefined. As with the function in Example 7, when the denominator approaches 0, the function value nears either positive or negative infinity.

The graph of $f(x) = \dfrac{6}{x-3}$ has an unmarked horizontal asymptote at the x-axis, or where $f(x) = 0$. Although it approaches 0, $f(x)$ never actually reaches the x-axis. The graph of $g(x) = \dfrac{8x}{\left(x-3\right)\left(x+6\right)}$ also has a horizontal asymptote at the x-axis for the left and right extensions of its graph, because the greater degree of x in the denominator means that the denominator will increase at a much greater rate than the numerator as x becomes very large.

However, unlike $f(x)$, the graph of $g(x)$ does have one x-intercept, at the origin. Remember, the x-intercept of a function occurs when the function is equal to 0. This happens for $g(x)$ when $x = 0$, because the numerator, $8x$, is then equal to 0, while the denominator is not. The function $f(x)$ will never have a value of 0, because its numerator, 6, will never equal 0.

Now you know that a zero of only the numerator in a rational function corresponds to an x-intercept, while a zero in the denominator corresponds to a discontinuity, such as a vertical asymptote. What happens when the numerator and denominator have a common zero? Let's take a look.

It is significant that the denominator of $\dfrac{8x}{x^2+3x-18}$ does **not** equal 0 when the numerator equals 0 (when $x = 0$). This means that $g(x)$ has a value of 0 at this point. If the denominator were also equal to 0, then it would be undefined, because 0/0 is undefined.

Graph the function $h(x) = \dfrac{x^2-1}{x^2-6x+5}$.

First, let's factor both the numerator and the denominator and cancel out any common factors. Remember, when you cancel out a common factor in a rational expression, you must note that x cannot equal the value that would make the factor equal to 0.

$$h(x) = \frac{(x+1)(x-1)}{(x-5)(x-1)}$$

$$h(x) = \frac{x+1}{x-5} \text{ where } x \neq 1$$

In Lesson 2.4, as we inspected and simplified rational expressions, we noted when an *x*-value was excluded from the domain for making a rational expression undefined. As we look at rational functions, these domain restrictions translate to certain characteristics of the function graphs, such as vertical asymptotes.

The graph has an *x*-intercept where $x + 1 = 0$, or when $x = -1$. The graph has a *y*-intercept where $x = 0$, or at $h(0) = \dfrac{0+1}{0-5} = -1/5$. The graph has a vertical asymptote where $x - 5 = 0$, or when $x = 5$. We can also create a table of values using the simplified equation, as long as we remember that $h(x)$ is undefined when $x = 1$.

x	−1	0	2	3	4	6	7	8	11
$h(x)$	0	$-\dfrac{1}{5}$	−1	−2	−5	7	4	3	2

Here is a graph of $h(x) = \dfrac{x+1}{x-5}$ where $x \neq 1$.

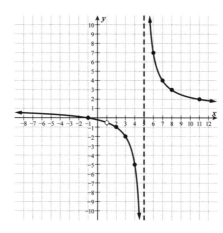

Unlike in the previous two examples, this graph does not have a horizontal asymptote at $y = 0$. Instead, the horizontal asymptote for $y = \dfrac{x+1}{x-5}$ and for $y = \dfrac{x^2-1}{x^2-6x+5}$ is at $y = 1$. This is because, for very large values of x, only the x terms to the greatest degree influence the y-values in any significant way. When x is huge, x^2 is much greater than $6x$, and any x term is definitely much greater than any constant. An x^3 term would have much greater influence than any x^2 term.

When the numerator and denominator are both polynomials to the same degree, then only the highest degree terms have much effect near infinity or negative infinity. This means that for those extreme values of x, the value of $\dfrac{x^2-1}{x^2-6x+5}$ becomes close to $\dfrac{x^2}{x^2}$, which simplifies to 1. If one or both of the x^2 terms had other coefficients, then the y-values would approach the ratio of those coefficients near x-values of ∞ or $-\infty$.

In Examples 7 and 8, we looked at how increasing values of x result in different levels of increase for x terms and x^2 terms, while constants remain unchanged. These differences of scale explain why a rational expression with same degree polynomials will tend toward just the ratio of its highest degree terms.

When the numerator and denominator polynomials of a rational function are of the same degree, the horizontal asymptote is the line $y = \dfrac{\text{leading coefficient of numerator}}{\text{leading coefficient of denominator}}$. In Example 9, $y = 1/1 = 1$.

When the denominator polynomial is of a greater degree than the numerator, the horizontal asymptote is the x-axis, as in the functions in Example 8.

When the numerator polynomial is of a greater degree than the denominator, there is no horizontal asymptote.

When the numerator polynomial is of a greater degree than the denominator, then as x approaches ∞, the function value will approach either positive or negative ∞, because the absolute value of the numerator will increase much more quickly than that of the denominator.

There is a hole in the graph at $(1, -1/2)$ because $h(x) = \dfrac{x+1}{x-5}$ would have a value of $-1/2$ when $x = 1$, except that we established that x cannot equal 1.

Try graphing the original rational function, $h(x) = \dfrac{x^2-1}{x^2-6x+5}$ using graphing technology. It looks exactly like our graph of $h(x) = \dfrac{x+1}{x-5}$ where $x \neq 1$.

> Any zero of the denominator in a rational function expression corresponds to a discontinuity in the function's graph at that x-value. If the zero is also a zero of the numerator, then the discontinuity is a hole in the graph. If the zero is not a common zero for the numerator, then the discontinuity is a vertical asymptote.

We need to acknowledge any discontinuities when describing the domain or range of a function. The domain of $h(x) = \dfrac{x^2 - 1}{x^2 - 6x + 5}$ is all real numbers except for 1 and 5, and the range is all real numbers except for $-1/2$ and 1.

USING SYSTEMS OF RATIONAL FUNCTIONS TO SOLVE RATIONAL EQUATIONS

Another way to solve a rational equation is to rewrite it as a system of functions, graph the functions, and find the x-values of the points of intersection.

EXAMPLE ════════════════════════════════════

Find all solutions to the following equation.

$$\frac{3x + 12}{x - 2} = \frac{x^2 - 10x + 16}{x - 8}$$

We can create the following system of equations.

$$f(x) = \frac{3x + 12}{x - 2}$$

$$g(x) = \frac{x^2 - 10x + 16}{x - 8}$$

In Lesson 2.2, we explored how to use systems of equations to solve polynomial equations. The same principles apply when the given equation is a rational equation. The x-values of the points of intersection are the x-values that make the original equation true.

Any point where the graphs of $f(x)$ and $g(x)$ intersect has the same x-value and the same function value [$f(x) = g(x)$]. The x-values of any points of intersection are the solutions to the equation $\dfrac{3x+12}{x-2} = \dfrac{x^2-10x+16}{x-8}$.

From its equation, we know that the graph of $f(x) = \dfrac{3x+12}{x-2}$ has a y-intercept of −6, an x-intercept of −4, a vertical asymptote at $x = 2$, and a horizontal asymptote at $y = 3$. It is always decreasing over each of its continuous sections. Here is its graph.

The y-intercept occurs when $x = 0$. The x-intercept occurs when y, or $f(x)$, equals 0, which is when the numerator, alone, of the rational expression equals 0. A vertical asymptote occurs when the denominator, alone, of the rational expression equals 0. For a rational function with same degree polynomials in the numerator and denominator, there is a horizontal asymptote at the level of the ratio of the leading coefficients.

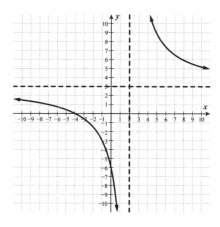

The function $g(x) = \dfrac{x^2-10x+16}{x-8}$ can be rewritten as $g(x) = \dfrac{\left(x-2\right)\left(x-8\right)}{x-8}$. The numerator and denominator share a common factor of $(x - 8)$, which we can cancel out as long as we also note that $x - 8$ cannot equal 0. So, $g(x) = x - 2$ where $x \neq 8$. The graph of $g(x)$ is the straight line $y = x - 2$ with a hole at the point (8, 6). Let's graph $g(x)$ on the same coordinate grid as $f(x)$.

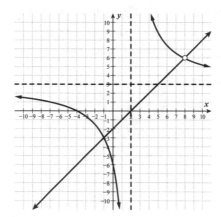

The two functions would have intersected at (8, 6), but there is a hole in the graph of $g(x)$ at this point. Since $g(x)$ does not exist when $x = 8$, 8 is not a solution for $f(x) = g(x)$. However, the two graphs do intersect at (−1, −3), where there is no discontinuity. The only solution to the original equation is $x = −1$. Let's test it, just to make sure.

$$\frac{3(-1)+12}{-1-2} \overset{?}{=} \frac{(-1)^2 - 10(-1)+16}{-1-8}$$

$$9/-3 \overset{?}{=} 27/-9$$

$$-3 = -3$$

The equation is true, so −1 is the solution for x. Notice that if we had tested 8 in the original equation, it would have produced a denominator of 0 in one fraction, which is undefined.

RATIONAL FUNCTIONS IN THE REAL WORLD

With knowledge of how key features of a function are related to its graph and to its equation, you can compare functions that are described or defined in different ways.

Two music venues each sold out quickly for their respective most popular concerts of the year. Venue A's most popular concert had a ratio, r, of available seats to unavailable seats given by the equation $r = \dfrac{200-10m}{5m+1}$, where m was the number of minutes since the first ticket was sold. Venue B had a ratio, r, of available seats to unavailable seats m minutes after the first ticket was sold as shown in the function table below. In this table, the m-value of 15 represents the lowest m-value for which $r = 0$.

m	r
0	150
1	35
3	12
8	2.8
13	0.5
15	0

Which venue's seat ratio function has a greater r-intercept, and what does that represent? Which venue's seat ratio function has a greater m-intercept, and what does that represent?

The r-intercept occurs when $m = 0$, so the table shows that Venue B has an r-intercept of 150. To find the r-intercept for Venue A, substitute 0 for m and solve for r.

$$r = \frac{200-10\left(0\right)}{5\left(0\right)+1} = 200/1 = 200$$

Venue A has an r-intercept of 200, which is greater than Venue B's r-intercept of 150. The r-value is the ratio of available seats to unavailable seats m minutes after the first ticket was sold. When $m = 0$, the very first ticket has just been purchased, so there is only one unavailable seat. At this point, Venue A has a ratio of 200/1, or 200 available seats to 1 unavailable seat, and Venue B has a ratio of 150/1, or 150 available seats to 1 unavailable seat. This means that Venue A must have a total of 201 seats and Venue B must have 151 seats. Venue A has greater seating capacity.

The m-intercept occurs when $r = 0$. We can see from the table that Venue B has an m-intercept of 15. To find the m-intercept for Venue A, substitute 0 for r and solve for m.

$$0 = \frac{200 - 10m}{5m + 1}$$

$0 = 200 - 10m$ where $m \neq -1/5$

$10m = 200$

$m = 20$

Time can't be negative for this situation anyway, so we don't have to worry about the fact that m specifically can't equal $-1/5$.

Venue A has an m-intercept of 20, which is greater than Venue B's m-intercept of 15. When $r = 0$, the ratio of available to unavailable seats is zero, so there are no available seats. This is the point at which each venue sells out of tickets to its concert. The variable m represents the minutes since the tickets went on sale. Venue A sold all tickets to its most popular concert in 20 minutes, and Venue B sold all tickets to its most popular concert in 15 minutes. It took Venue A longer to sell out than Venue B.

In Example 11, we were not given the equation for Venue B, but in situations where you are given both of the two equations—or can write them from given information—graphing can be very useful for comparing rational functions.

EXAMPLE 12

Karl can sand 4 square meters of wood floor in x hours. Dawn can sand at a rate that is 1 square meter per hour faster than 40% of Karl's rate. Is one of them definitely faster than the other? If x were defined, how would that answer the question of who sands faster? Explain for which x-values Karl would be faster and for which x-values Dawn would be faster.

First, let's write functions relating each person's rate to time. Karl can sand 4 square meters in x hours, so $k = 4/x$, where k represents Karl's sanding rate, in square meters per hour. Dawn's rate is 1 square meter per hour faster than 40% of Karl's rate, so $d = 0.4(4/x) + 1$, where d represents Dawn's sanding rate, in square meters per hour. This equation can be rewritten as the rational function $d = \dfrac{x + 1.6}{x}$.

Let's graph each of these functions on the same coordinate grid to see how the rates *k* and *d* relate to *x* and to one another.

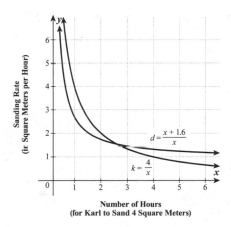

Number of Hours
(for Karl to Sand 4 Square Meters)

The two functions intersect somewhere between *x* = 2 and *x* = 3, so neither Karl nor Dawn is faster for the entire set of *x*-values. To the left of the point of intersection, *k* is greater than *d*, and to the right of it, *d* is greater than *k*. Let's use the equations to find the exact *x*-value of the point of intersection, where Karl and Dawn would have the same sanding rate.

$\dfrac{4}{x} = \dfrac{x+1.6}{x}$	Note that $x \neq 0$.
$4x = x^2 + 1.6x$	Cross-multiply.
$0 = x^2 - 2.4x$	Subtract $4x$ from both sides.
$0 = x(x - 2.4)$	Factor the quadratic.
$0 = x$ or $0 = x - 2.4$	Set each factor equal to 0.

Because $x \neq 0$, 0 is an extraneous solution, and the only true solution is $x = 2.4$. Karl and Dawn have the same sanding rate if Karl sands 4 square meters in exactly 2.4 hours.

If *x*, the number of hours it takes Karl to sand 4 square meters, is defined as less than 2.4 hours, then Karl is the faster sander. If *x* is defined as greater than 2.4 hours, then Dawn is the faster sander.

Lesson 3.4
Radical Equations in One Variable

An **inverse operation** undoes what the original operation did. Addition and subtraction are inverse operations, and multiplication and division are inverse operations. Taking the square root is the inverse operation of squaring, and taking the cube root is the inverse operation of cubing.

A **radical equation** is an equation that contains a radical expression. The radicals you will encounter most often are square roots, but radicals also include cube roots and roots of other degrees.

How does squaring both sides produce extraneous solutions? Look at a numerical example. The equation $-3 = 3$ is false, but when you square both sides, you get the true equation $9 = 9$. When we find solutions to a squared equation, they may not actually make the original radical equation true.

> To solve a radical equation, isolate the radical on one side of the equation and use the inverse operation. For example, for a square root, square both sides, and for a cube root, cube both sides. Just watch out for extraneous solutions when raising to an even power!

EXAMPLE

Solve the following equation for x.

$$\sqrt{x-1} + 3 = x$$

First, subtract 3 from both sides, to isolate the radical expression.

$\sqrt{x-1} = x - 3$

$x - 1 = (x - 3)^2$ Square both sides of the equation.

$x - 1 = x^2 - 6x + 9$ Expand the squared binomial.

$0 = x^2 - 7x + 10$ Move all terms to one side of the equation.

$0 = (x - 5)(x - 2)$ Factor the quadratic.

$x = 5$ or $x = 2$ Solve for each factor set equal to 0.

Test each of these values in the original equation. We cannot use any other equation from later in the solution process, because the act of squaring is what produces extraneous solutions, and squaring both sides was one of the first things we did.

$$\sqrt{5-1} + 3 \overset{?}{=} 5 \qquad\qquad \sqrt{2-1} + 3 \overset{?}{=} 2$$

$$\sqrt{4} + 3 \overset{?}{=} 5 \qquad\qquad\quad \sqrt{1} + 3 \overset{?}{=} 2$$

$$2 + 3 \overset{?}{=} 5 \qquad\qquad\quad 1 + 3 \overset{?}{=} 2$$

$$5 = 5 \qquad\qquad\qquad 4 \neq 2$$

The equation $5 = 5$ is true, so $x = 5$ is a solution. The equation $4 = 2$ is not true, so $x = 2$ is an extraneous solution. The only solution to the equation $\sqrt{x-1} + 3 = x$ is $x = 5$.

EXAMPLE

Find all possible solutions for *a*.

$$\sqrt[3]{a^2 + 4a - 4} - a = 0$$

Add *a* to both sides to isolate the cube root on one side of the equation.

$\sqrt[3]{a^2 + 4a - 4} = a$	
$a^2 + 4a - 4 = a^3$	Cube both sides.
$0 = a^3 - a^2 - 4a + 4$	Move all terms to one side of the equation.
$0 = a^2(a - 1) - 4(a - 1)$	Factor out the common factor from each pair of terms.
$0 = (a - 1)(a^2 - 4)$	Use the distributive property to rewrite as two factors.
$0 = (a - 1)(a + 2)(a - 2)$	Factor the difference of squares.

If you didn't see the trick of factoring the cubic into two pairs of (*a* – 1) multiplied by some common factor, you could have used the Rational Root Theorem and Remainder Theorem to find the full factorization. For a review of these theorems and their application to factoring polynomials, see Chapter 1.

Notice the lack of extraneous solutions? Cubing a number preserves its sign, whether negative or positive, so cubing does not produce extraneous solutions the way squaring or raising to another even-degree power might.

The potential solutions are $a = 1$, $a = -2$, and $a = 2$. Let's test each one in the original equation.

$$\sqrt[3]{1^2 + 4(1) - 4} - 1 \overset{?}{=} 0$$

$$\sqrt[3]{1} - 1 \overset{?}{=} 0$$

$$1 - 1 \overset{?}{=} 0$$

$$0 = 0$$

$$\sqrt[3]{(-2)^2 + 4(-2) - 4} - (-2) \overset{?}{=} 0$$

$$\sqrt[3]{-8} + 2 \overset{?}{=} 0$$

$$-2 + 2 \overset{?}{=} 0$$

$$0 = 0$$

$$\sqrt[3]{2^2 + 4(2) - 4} - 2 \overset{?}{=} 0$$

$$\sqrt[3]{8} - 2 \overset{?}{=} 0$$

$$2 - 2 \overset{?}{=} 0$$

$$0 = 0$$

All three a-values produce true numerical equations, so all three are possible solutions. The possible solutions are −2, 1, and 2.

Here is how you may see radical equations on the ACT.

If $\sqrt{2y^2 - 17} = y + 2$, then what could be the value of y?

   A. −3 and 7 only

   B. 3 and 7 only

   C. −3 only

   D. 3 only

   E. 7 only

Lesson 3.5
Radical Inequalities in One Variable

The solution process for radical inequalities is similar to the solution process for rational inequalities.

> To solve a radical inequality, solve an equation version of the inequality and use the potential solutions as endpoints of intervals to test in the original inequality. Also, neither the radicand nor the value of the radical can be negative for square roots, so use the variable value that makes the radicand 0 as another endpoint demarcating another interval to test.

These rules are for finding all real number solutions. With imaginary numbers, the radicand could be negative.

The **radicand** is the value inside the radical symbol.

EXAMPLE

Solve the following inequality for x.

$$\sqrt{2x+8} \geq x$$

The radical is a square root, so we know that the radicand must be greater than or equal to 0, so $2x + 8 \geq 0$, which simplifies to $x \geq -4$. Keeping this limitation in mind, let's solve the inequality as an equation.

$\sqrt{2x+8} = x$	Write as an equation.
$2x + 8 = x^2$	Square both sides.
$0 = x^2 - 2x - 8$	Move all terms to one side of the equation.
$0 = (x + 2)(x - 4)$	Factor the quadratic.
$x = -2$ or $x = 4$	Solve each factor equal to 0.

Let's plot the points -2 and 4 on a number line. Also, let's plot the point -4, because we know that $x \geq -4$.

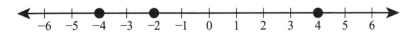

Now we'll test points in each interval.

When $x = -5$: $\sqrt{2(-5)+8}$ is equal to $\sqrt{-2}$, so this is not a solution in the set of real numbers.

When $x = -3$: $\sqrt{2(-3)+8} \geq -3$ is true, because $\sqrt{2} \geq -3$ is true.

When $x = 0$: $\sqrt{2(0)+8} \geq 0$ is true, because $\sqrt{8} \geq 0$ is true.

When $x = 5$: $\sqrt{2(5)+8} \geq 5$ is false, because $\sqrt{18}$ is not greater than or equal to 5.

The solution set is the interval between −4 and −2 and the interval between −2 and 4. Let's also test each of the endpoints of the intervals.

When $x = -4$: $\sqrt{2(-4)+8} \geq -4$ is true, because $0 \geq -4$ is true.

When $x = -2$: $\sqrt{2(-2)+8} \geq -2$ is true, because $2 \geq -2$ is true.

When $x = 4$: $\sqrt{2(4)+8} \geq 2$ is true, because $4 \geq 2$ is true.

ACT A

The radical is already isolated on the left side of the equation, so we can immediately square both sides of the equation.

$2y^2 - 17 = (y + 2)^2$

$2y^2 - 17 = y^2 + 4y + 4$ Expand the squared binomial.

$y^2 - 4y - 21 = 0$ Move all terms to one side of the equation.

$(y + 3)(y - 7) = 0$ Factor the quadratic.

$y = -3$ or $y = 7$ Solve for each factor equal to 0.

Hold on! Before you mark (A), test each value in the original equation.

$\sqrt{2(-3)^2 - 17} \stackrel{?}{=} -3 + 2$ $\sqrt{2(7^2) - 17} \stackrel{?}{=} 7 + 2$

$\sqrt{18 - 17} \stackrel{?}{=} -1$ $\sqrt{98 - 17} \stackrel{?}{=} 9$

$\sqrt{1} \stackrel{?}{=} -1$ $\sqrt{81} \stackrel{?}{=} 9$

$1 \neq -1$ $9 = 9$

The equation 1 = −1 is not true, so −3 is an extraneous solution. The equation 9 = 9 is true, so 7 is the only solution for y. The answer is (E).

The points −4, −2, and 4 are all included in the solution set. The solution set is shown on the number line below.

The solution set can also be written as $-4 \le x \le 4$.

We no longer need to mark a closed-circle point at −2, because it is not an endpoint of a solution set. It is simply included in the solution set, along with all other values between −4 and 4.

Another way to solve Example 15 is by creating and graphing a system of equations. In this case, instead of just looking for points of intersection, we will look at the graph to see where the value of $\sqrt{2x+8}$ is greater than or equal to the value of x. The system of equations is shown below.

$$y = \sqrt{2x+8}$$

$$y = x$$

Using graphing technology, we can graph the two functions on the same coordinate grid, as shown below.

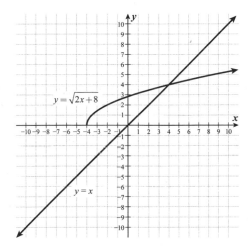

We are using a method similar to the one used in Example 10, of graphing a system of equations to relate the values of the two sides of the given equation. However, instead of finding points with the **same** function value, here we are comparing to see where $\sqrt{2x+8}$ has an equal **or greater** value than x.

If we were solving the inequality $\sqrt{2x+8} > x$, then the solution set would be $-4 \le x < 4$. The x-value of 4 makes $\sqrt{2x+8}$ equal to x, so it is included in the solution when the inequality symbol is $\ge$ but not when the inequality symbol is $>$.

The two graphs intersect at the point (4, 4). The graph of $y = \sqrt{2x+8}$ only exists for $x \ge -4$. The graph of $y = \sqrt{2x+8}$ is higher than the graph of $y = x$ from $x = -4$ to $x = 4$, where they intersect. So, $\sqrt{2x+8} \ge x$ for $-4 \le x \le 4$.

Lesson 3.6
Radical Functions

The graph of a quadratic function is a parabola. It opens upward if the coefficient of x^2 is positive and opens downward if the coefficient of x^2 is negative.

If a constant is added to a function value, the function graph is translated vertically—up if the constant is positive, down if the constant is negative.

If a constant is added to every x-value in a function, the function graph is translated horizontally—to the left if the constant is positive, to the right if the constant is negative.

The **Pythagorean Theorem** states that the sum of the squares of the leg lengths of a right triangle is equal to the square of its hypotenuse: $a^2 + b^2 = c^2$.

A **radical function** is a function that is defined as a radical expression.

In this lesson, we will only explore square root functions.

EXAMPLE 16

Graph the function $y = \sqrt{x+1}$.

If we square both sides, we get $y^2 = x + 1$, which we can rewrite as $x = y^2 - 1$. This looks like a familiar quadratic equation, but it tells you how x is related to y, instead of how y is related to x. Remember how you graph quadratic functions.

The quadratic $y = x^2 - 1$ looks like this.

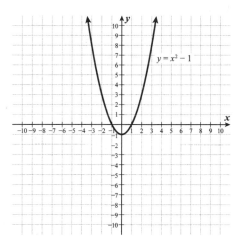

When the variables are switched, as in $x = y^2 - 1$, the relationship produces the same parabola but turned 90° clockwise. Actually, it is technically a reflection, not a rotation. When a function has its x and y variables switched, it gets reflected across the line $y = x$. The result for this parabola is the same, though. The graph of $x = y^2 - 1$ is shown below.

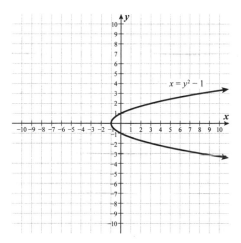

In the function $y = \sqrt{x+1}$, the radicand cannot be negative, so $x + 1 \geq 0$, which means that $x \geq -1$. This is already evident in the graph of $x = y^2 - 1$.

However, the function $y = \sqrt{x+1}$ is not exactly the same as $x = y^2 - 1$. We know that the value of a square root must be greater than or equal to 0, so $\sqrt{x+1} \geq 0$, or $y \geq 0$. The half of the parabola that is below the x-axis represents negative y-values, so it is not included in the function $y = \sqrt{x+1}$. The graph of $y = \sqrt{x+1}$ is the half of the parabola $x = y^2 - 1$ that is above the x-axis.

The point $(-1, 0)$ on the x-axis is included in the function, because a square root **can** equal 0 ($\sqrt{0} = 0$). The square root just can't be negative.

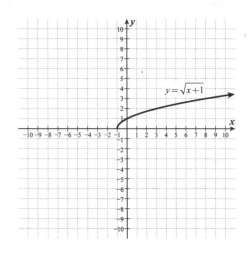

The domain of the function $y = \sqrt{x+1}$ is $x \geq -1$, as you can see from the graph (starting at $x = -1$ and continuing indefinitely to the right), as well as from the equation ($x + 1 \geq 0$, because the radicand must be non-negative). The range of this function is $y \geq 0$, as you can see from the graph (starting at $y = 0$ and continuing indefinitely upward), as well as from the equation (a square root, such as $\sqrt{x+1}$, must be 0 or any positive number).

> The graph of a basic square root function in the form
> $f(x) = \sqrt{x-a}$, where a is a constant, is half of a parabola.

Another way to graph the function $y = \sqrt{x+1}$ is to graph it as a translation of its parent function, $y = \sqrt{x}$.

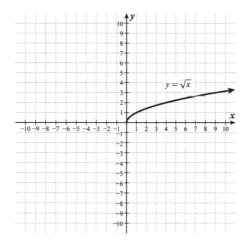

In Lesson 1.3, you saw how changes to x or to the function value translate a given polynomial function. These rules apply to any kind of function, including radical functions.

For any kind of function, including radical functions, the following translation rules apply for a function $f(x)$ and constant k.

The graph of $f(x) + k$ represents a vertical shift of the graph of $f(x)$ by k units—up if k is positive, down if k is negative.

The graph of $f(x + k)$ represents a horizontal shift of the graph of $f(x)$ by k units—to the left if k is positive, to the right if k is negative.

In the case of $y = \sqrt{x+1}$, a constant of 1 is added to x, inside the radicand, so the graph of $y = \sqrt{x+1}$ is the graph of $y = \sqrt{x}$ shifted 1 unit to the left.

Graph the function $h(x) = \sqrt{x-4} - 7$.

The parent function is $y = \sqrt{x}$. The constant 4 is subtracted from x, inside the radicand, so the graph of $y = \sqrt{x}$ will be shifted 4 units to the right. The constant 7 is subtracted from the radical value, so the graph will be shifted 7 units down.

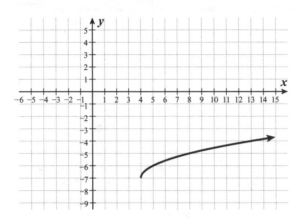

This is the graph of $h(x) = \sqrt{x-4} - 7$. To make sure, let's test values of x in the equation and see if the graph includes those points.

The radical $\sqrt{x-4}$ only exists as a real number when $x - 4 \geq 0$, or $x \geq 4$. That matches our graph, which starts at $x = 4$ and moves to the right.

When $x = 4$, $h(4) = \sqrt{4-4} - 7 = -7$.

When $x = 5$, $h(5) = \sqrt{5-4} - 7 = -6$.

When $x = 8$, $h(8) = \sqrt{8-4} - 7 = -5$.

When $x = 13$, $h(13) = \sqrt{13-4} - 7 = -4$.

The equation produces the coordinate pairs (4, –7), (5, –6), (8, –5), and (13, –4), and the graph passes through each of these points. We have correctly graphed $h(x) = \sqrt{x-4} - 7$.

The function $f(x) = \sqrt{2x+6}$ is not as easy to graph, because the parent function is not simply translated in this case. The magnitude of the half parabola is also affected, because x is multiplied by 2 in the radicand. However, we can still determine key characteristics of the graph of $f(x) = \sqrt{2x+6}$ by looking at its equation.

EXAMPLE

Graph the function $f(x) = \sqrt{2x+6}$.

The smallest possible value of any square root is always 0, so the minimum value of $f(x)$ is 0. This means that the lowest point on the graph of $f(x) = \sqrt{2x+6}$ is on the x-axis (where $f(x) = 0$) and the rest of the graph is above the x-axis.

Let's find the smallest possible value of x for this function. The value of the radicand must be greater than or equal to 0, so $2x + 6 \geq 0$.

$$2x \geq -6 \qquad\qquad \text{Subtract 6 from both sides.}$$
$$x \geq -3 \qquad\qquad \text{Divide both sides by 2.}$$

Since x-values for $f(x) = \sqrt{2x+6}$ are always greater than or equal to -3, we know that the graph starts at $x = -3$ and moves to the right. The half parabola begins at the point $(-3, 0)$ and curves up and to the right similarly to the $y = \sqrt{x}$ function. We just need to find points that the graph passes through, to complete our graph. Choose values of x that make $(2x + 6)$ a perfect square.

When $x = -5/2$, $f(-5/2) = \sqrt{2\left(-\dfrac{5}{2}\right) + 6} = \sqrt{1} = 1.$

When $x = -1$, $f(-1) = \sqrt{2(-1)+6} = \sqrt{4} = 2.$

When $x = 3/2$, $f(3/2) = \sqrt{2\left(\dfrac{3}{2}\right) + 6} = \sqrt{9} = 3.$

When $x = 5$, $f(5) = \sqrt{2(5)+6} = \sqrt{16} = 4.$

The graph begins at (−3, 0) and passes through (−5/2, 1), (−1, 2), (3/2, 3), and (5, 4).

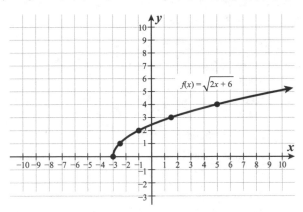

RADICAL FUNCTIONS IN THE REAL WORLD

Radical functions may be useful for solving problems where the relationship involves one or more squared variables, such as a relationship described by the Pythagorean theorem.

EXAMPLE

> **Yimin is decorating a series of cards of various sizes by adding ribbons along their diagonals. Each card is a rectangle that is 7 centimeters longer than it is wide. Write a formula for the length of ribbon needed for a rectangular card in terms of the card's width. If one card has a diagonal that is 3 centimeters more than twice the width of the rectangle, what are the dimensions of this rectangular card?**
>
> Let w = the width of the rectangular card, l = the length of the card, and d = the length of the diagonal of the card. The length is 7 centimeters longer than the width, so $l = w + 7$.

It's often helpful to draw diagrams to visualize geometric relationships.

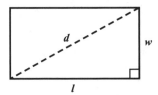

As you can see from the diagram, the width, length, and diagonal of a rectangle form a right triangle, so we can write an equation using the Pythagorean theorem.

$l^2 + w^2 = d^2$

$d = \sqrt{l^2 + w^2}$	Solve for d.
$d = \sqrt{(w+7)^2 + w^2}$	Substitute $(w + 7)$ for l.
$d = \sqrt{w^2 + 14w + 49 + w^2}$	Expand the squared binomial in the radicand.
$d = \sqrt{2w^2 + 14w + 49}$	Combine like terms.

This formula gives the length, d, of ribbon needed for a card of width w in Yimin's collection.

If the diagonal of a card is 3 centimeters more than twice the width, then $d = 2w + 3$. We can substitute $(2w + 3)$ for d in our radical function equation.

$2w + 3 = \sqrt{2w^2 + 14w + 49}$	
$(2w + 3)^2 = 2w^2 + 14w + 49$	Square both sides.
$4w^2 + 12w + 9 = 2w^2 + 14w + 49$	Expand the squared binomial.
$2w^2 - 2w - 40 = 0$	Move all terms to one side.
$w^2 - w - 20 = 0$	Divide both sides by 2.
$(w + 4)(w - 5) = 0$	Factor the quadratic.

The solutions to the equation are $w = -4$ and $w = 5$. However, a card cannot have a negative length for either side length, so the width cannot equal -4. This is an extraneous solution. In fact, if you test $w = -4$ in the equation $2w + 3 = \sqrt{2w^2 + 14w + 49}$, you will get a false numerical equation ($-5 = 5$).

The solution is $w = 5$. The width of the card is 5 centimeters. The length of the card is $5 + 7 = 12$ centimeters. The diagonal of the card, or the length of the ribbon needed for this card, is $2(5) + 3 = 13$ centimeters.

We could also solve by graphing a system of equations, after correctly translating the given information into equations. Here is our system of equations relating the variables *d* and *w*.

Notice that the radical function is a different shape from most of the ones we have looked at so far. Having an x^2 (or w^2, in this case) within the radicand produces a very different graph. Graphing technology is great for these situations!

$$d = \sqrt{2w^2 + 14w + 49}$$

$$d = 2w + 3$$

Using graphing technology, we can graph both functions on the same coordinate grid.

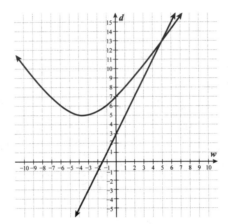

The two graphs intersect at the point (5, 13), so the solution to the system of equations is *w* = 5, as we found algebraically. The width of the card is 5 centimeters. The *d*-value of this point of intersection is 13, so the length of the diagonal, the length of the ribbon, is 13 centimeters. The length of the card is 7 centimeters more than its width, or 12 centimeters.

CHAPTER 3 PRACTICE QUESTIONS

Directions: Complete the following open-ended problems as specified by each question stem. For extra practice after answering each question, try using an alternative method to solve the problem or check your work.

1. Solve the equation and check your solution.

$$\frac{x}{x-3} + \frac{2x}{x+3} = \frac{18}{x^2-9}$$

2. Sketch the graph of $f(x) = \dfrac{2\left(x^2-9\right)}{x^2-4}$ by discussing asymptotes, intercepts, symmetry, and finding a couple of additional points to plot. Use a calculator to check your graph.

3. Find all real solutions of $\dfrac{1}{\left(x+1\right)^2} = \dfrac{1}{x+1} + 2.$

4. Find all real solutions to $\sqrt[4]{5x^2-6} = x.$

5. Solve $\sqrt{x} - 1 = \sqrt{2x+1}$ and check the solution(s).

6. Solve $\sqrt{x+1} < 2$ and check the solution.

7. The length of a rectangle is 7 units more than its width. Find the range of values for the length and the width if the ratio of the length to the width is no more than 10 and at least 5.

8. On her first ski run this morning, Mitsue skied a 0.3-mile trail. The second trail she skied was twice as long and steeper, and her speed on it was 1.5 times as fast as her speed on the first trail. Her ski time for the second trail was half a minute more than her time for the first trail. Write an equation for t, Mitsue's time on the first trail, as a function of r, her skiing speed on the first trail. Write another equation to represent her second trail run, in terms of the same values of t and r. What was Mitsue's skiing speed, in miles per hour, for each of the two trails?

9. Jean-Luc needs to install a new cylindrical storage tank water heater. To minimize how much space it takes in his basement, he plans to get the tallest one possible that will fit, which is 5 feet 3 inches high. Write a function for r, the radius, in feet, of a storage tank of height 5 feet 3 inches, in terms of v, the volume of water, in cubic feet, that it holds. Jean-Luc decides to get an 80-gallon storage tank. If one cubic foot contains 7.48 gallons, what is the **diameter** of the storage tank he chose?

SOLUTIONS TO CHAPTER 3 PRACTICE QUESTIONS

1. **−2**

    Multiply each side by the lowest common denominator of $(x + 3)(x − 3)$:

    $$\frac{x}{x-3}(x+3)(x-3) + \frac{2x}{x+3}(x+3)(x-3) = \frac{18}{x^2-9}(x+3)(x-3)$$

    $x(x + 3) + 2x(x − 3) = 18$
    $x^2 + 3x + 2x^2 − 6x = 18$
    $3x^2 − 3x = 18$
    $3x^2 − 3x − 18 = 0$
    $3(x^2 − x − 6) = 0$
    $3(x − 3)(x + 2) = 0$
    $x = 3 \; ; \; x = −2$

    Don't stop here! Now, check the solutions in the original equation. If 3 is plugged back in for x, then there is a zero in the denominator of two expressions, which means that 3 is an extraneous solution. So, check -2:

    $$\frac{x}{x-3} + \frac{2x}{x+3} = \frac{18}{x^2-9}$$

    $$\frac{(-2)}{(-2)-3} + \frac{2(-2)}{(-2)+3} = \frac{18}{(-2)^2-9}$$

    $$\frac{2}{5} - \frac{4}{1} = \frac{18}{-5}$$

    $$\frac{2}{5} - \frac{20}{5} = -\frac{18}{5}$$

    The statement is true, so the solution is $x = −2$.

2.

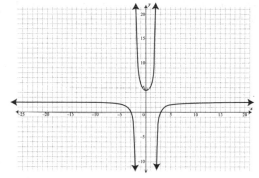

To sketch the graph, begin by finding the x- and y-intercepts. Factor the numerator and denominator of the given equation:

$$f(x) = \frac{2(x^2-9)}{x^2-4} = \frac{2(x+3)(x-3)}{(x+2)(x-2)}$$

To find the y-intercept, set $x = 0$:

$$\frac{2(x+3)(x-3)}{(x+2)(x-2)} = \frac{2((0)+3)((0)-3)}{((0)+2)((0)-2)} = \frac{2(3)(-3)}{(2)(-2)} = \frac{-18}{-4} = \frac{9}{2}$$

So, the y-intercept is at $(0, \frac{9}{2})$.

To find the x-intercept(s), set $y = 0$ and solve for x-values that make the numerator equal 0:

$0 = 2(x + 3)(x - 3)$
$0 = x + 3 \ ; \ \ 0 = x - 3$
$-3 = x \ ; \ \ 3 = x$

So, the x-intercepts are at $(\pm 3, 0)$.

Next, proceed to find the vertical and horizontal asymptotes. Find the vertical asymptote(s) by setting the denominator equal to zero and solving:

$(x + 2)(x - 2) = 0$
$x + 2 = 0 \ ; \ \ x - 2 = 0$
$x = -2 \ ; \ \ x = 2$

So, the vertical asymptotes occur at $x = \pm 2$.

Find the horizontal asymptote by comparing the degrees of the polynomials in the numerator and the denominator. Since the degrees are equal to each other (degree 2), look to the leading coefficients:

$$\frac{2}{1} = 2$$

So, the horizontal asymptote is at $y = 2$.

Check for symmetry with respect to the y-axis by finding $f(-x)$:

$$f(x) = \frac{2(x^2-9)}{x^2-4}$$

$$f(-x) = \frac{2\left((-x)^2-9\right)}{(-x)^2-4} = \frac{2(x^2-9)}{x^2-4}$$

Since $f(-x) = f(x)$, the function is even and has y-axis symmetry.

Plug in a couple of test values, such as $x = 1$ and $x = 4$, to see the behavior:

$$f(1) = \frac{2\left((1)^2-9\right)}{(1)^2-4} = \frac{2(-8)}{-3} = \frac{16}{3}$$

$$f(4) = \frac{2\left((4)^2-9\right)}{(4)^2-4} = \frac{2(7)}{12} = \frac{7}{6}$$

Finally, use the information gathered to sketch the graph.

3. **$x = -2, -1/2$**

The least common denominator of the three terms in this equation is $(x + 1)^2$.

$$\frac{1}{(x+1)^2} = \frac{1}{x+1}\left(\frac{x+1}{x+1}\right) + 2\left(\frac{(x+1)^2}{(x+1)^2}\right)$$

Convert each term to have the lowest common denominator.

$$\frac{1}{(x+1)^2} = \frac{x+1}{(x+1)^2} + \frac{2(x+1)^2}{(x+1)^2}$$

Since all denominators are the same, set the numerators of the left and right sides equal to each other. Note that $x \neq -1$ since the denominator will equal zero at that x-value.

$1 = (x + 1) + 2(x + 1)^2, x \neq -1$
$1 = x + 1 + 2(x^2 + 2x + 1)$ Expand the squared binomial using FOIL.
$1 = x + 1 + 2x^2 + 4x + 2$ Distribute 2 to each term in parentheses.
$0 = 2x^2 + 5x + 2$ Subtract 1 from both sides and combine like terms.

Next, to determine the solutions, factor the quadratic and set each term equal to zero. (The quadratic formula could also be used with $a = 2$, $b = 5$, and $c = 2$.)

$0 = 2x^2 + 5x + 2$
$0 = (2x + 1)(x + 2)$
$2x + 1 = 0$; $x + 2 = 0$
$x = -1/2$; $x = -2$

The solutions are $x = -2$ and $x = -1/2$.

4. $x = \sqrt{3}, \sqrt{2}$

To tackle this problem, clear out the radical by raising both sides to the fourth power:

$\sqrt[4]{5x^2 - 6} = x$

$(\sqrt[4]{5x^2 - 6})^4 = (x)^4$

$5x^2 - 6 = x^4$

Next, get the expression equal to zero by moving all terms to one side:

$5x^2 - 6 = x^4$
$0 = x^4 - 5x^2 + 6$

The terms are in decreasing order of exponents and can be factored:

$0 = x^4 - 5x^2 + 6$	Quadratic in form of $a^2 - 5a + 6$
$0 = (x^2 - 3)(x^2 - 2)$	Factored form of $(a - 3)(a - 2)$

Set each factor equal to zero and solve for x:

$0 = x^2 - 3$; $0 = x^2 - 2$	
$3 = x^2$	$2 = x^2$
$\pm\sqrt{3} = x$	$\pm\sqrt{2} = x$

Don't stop there! Remember to check your solutions by plugging back into the original equation.

The expression on the left-hand side of the equation is an even root, so it will always produce a positive value. The negative roots would produce a positive fourth-root expression on the left side of the equation and a negative expression on the right side (x), so they fail the check. So, check $x = \sqrt{3}$ and $\sqrt{2}$:

$\sqrt[4]{5x^2 - 6} = x$	$\sqrt[4]{5x^2 - 6} = x$
$\sqrt[4]{5(\sqrt{3})^2 - 6} = (\sqrt{3})$	$\sqrt[4]{5(\sqrt{2})^2 - 6} = (\sqrt{2})$

$$\sqrt[4]{5(3)-6} = \sqrt{3} \qquad\qquad \sqrt[4]{5(2)-6} = \sqrt{2}$$

$$\sqrt[4]{9} = \sqrt{3} \qquad\qquad \sqrt[4]{4} = \sqrt{2}$$

$$1.732 = 1.732 \qquad\qquad 1.414 = 1.414$$

Therefore, the solutions are $x = \sqrt{3}$ and $x = \sqrt{2}$.

5. **No solution.**
   Square both sides of the given equation to begin removing the square root:

$\sqrt{x} - 1 = \sqrt{2x+1}$	Given.
$(\sqrt{x} - 1)^2 = (\sqrt{2x+1})^2$	Square both sides.
$x - 2\sqrt{x} + 1 = 2x + 1$	FOIL the left-hand side.
$-2\sqrt{x} = x$	Combine like terms.

   Repeat the process of squaring both sides to remove the radical:

$-2\sqrt{x} = x$	
$(-2\sqrt{x})^2 = (x)^2$	Square both sides.
$4x = x^2$	
$0 = x^2 - 4x$	Subtract $4x$ from both sides.
$0 = x(x-4)$	Factor.
$0 = x \; ; \; 0 = x - 4$	Set each term equal to 0.
$0 = x \; ; \; 4 = x$	Solve for x.

   Don't stop there! Check both solutions!

$$\sqrt{x} - 1 = \sqrt{2x+1} \qquad\qquad \sqrt{x} - 1 = \sqrt{2x+1}$$

$$\sqrt{(0)} - 1 = \sqrt{2(0)+1} \qquad\qquad \sqrt{(4)} - 1 = \sqrt{2(4)+1}$$

$$-1 \neq 1 \qquad\qquad\qquad 1 \neq 3$$

Since both solutions failed the check, the equation $\sqrt{x} - 1 = \sqrt{2x+1}$ has no real solutions.

An alternative way to solve this equation is by graphing the system of equations $y = \sqrt{x} - 1$ and $y = \sqrt{2x+1}$, as shown below. Any solution to the equation $\sqrt{x} - 1 = \sqrt{2x+1}$ corresponds to an intersection of these two functions.

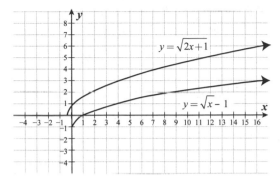

The two graphs never intersect. As x increases, the value of $\sqrt{2x+1}$ increases more quickly than the value of $\sqrt{x}-1$, so the graphs get farther and farther apart.

6. **$-1 \leq x < 3$**
    Begin by squaring both sides to remove the radical. This yields $x + 1 < 4$, which can be simplified to $x < 3$.

    Squaring both sides sometimes produces an incorrect solution set, so we must test values on either side of 3, to make sure that $x < 3$ is correct.

    When $x = 0$, the inequality becomes $\sqrt{0+1} < 2$, which simplifies to $1 < 2$. This is true.

    When $x = 8$, the inequality becomes $\sqrt{8+1} < 2$, which simplifies to $3 < 2$. This is false.

    The inequality $x < 3$ seems true. However, take a look back at the original function. Since there is a square root involved, the radicand cannot be negative. To finish out the solution, it must be determined where $x + 1 \geq 0$:

    $x + 1 \geq 0$
    $x \geq -1$

    So, the solution is $-1 \leq x < 3$.

7. **$70/9 \leq l \leq 35/4$ and $7/9 \leq w \leq 7/4$**
    First, determine expressions for the dimensions of the rectangle. Since the length is 7 more than the width, the width can be expressed as x and the length can be expressed as $x + 7$. Given that the ratio of the length to the width is at least 5 and no more than 10, an inequality can be set up as follows:

    $$5 \leq \frac{x+7}{x} \leq 10$$

    Begin solving the inequality by multiplying both all three expressions by x:

    $$5x \leq x + 7 \leq 10x$$

Solve by splitting the inequality up into two pieces:

$5x \leq x + 7$ and $x + 7 \leq 10x$
$4x \leq 7$ $7 \leq 9x$
$x \leq 7/4$ $7/9 \leq x$

So, the range of possible values for the width, x, is $7/9 \leq x \leq 7/4$. Since the length is 7 more than the width, the possible range for the length, $x + 7$, can be found by adding 7:

$7/9 + 7 \leq x + 7 \leq 7/4 + 7$

$70/9 \leq x + 7 \leq 35/4$

The dimensions of the rectangle are length l and width w given by $70/9 \leq l \leq 35/4$ and $7/9 \leq w \leq 7/4$.

8. **12 miles per hour, 18 miles per hour**
   This situation uses the distance = rate × time relationship. To write an equation for t as a function of r, rewrite the $d = rt$ equation as $t = d/r$. On Mitsue's first ski run this morning, she skied a distance of 0.3 mile, so the function $t = 0.3/r$ describes the run.

   On the second trail she skied, the distance was twice as long: $2 \cdot 0.3 = 0.6$ mile. Her speed was 1.5 times as fast: $1.5r$. Her time was half a minute more: $t + 0.5$. Use these new values in the $t = d/r$ relationship.

   $$t + 0.5 = \frac{0.6}{1.5r}$$

   Notice that t still represents Mistue's time, in minutes, on the first trail, and r still represents her speed on the first trail. Because these values are consistent, we can solve this system of two equations to find t and r.

$t = \dfrac{0.6}{1.5r} - 0.5$	Subtract 0.5 from both sides of the equation for the second trail.
$0.3/r = \dfrac{0.6}{1.5r} - 0.5$	Substitute the expression for t from the first trail equation, $t = 0.3/r$.
$0.45 = 0.6 - 0.75r$	Multiply both sides of the equation by the LCD, $1.5r$.
$-0.15 = -0.75r$	Subtract 0.6 from both sides.
$0.2 = r$	Divide both sides by -0.75.

   All values have been given in miles and minutes, so this solution means that Mitsue's speed on the first trail was 0.2 mile per minute. To convert to miles per hour, multiply by 60 minutes/ hour: $0.2 \cdot 60 = 12$. Mitsue's speed on the second trail was 1.5 times this speed: $1.5 \cdot 12 = 18$. Her speed on the first trail was 12 miles per hour, and her speed on the second trail was 18 miles per hour.

9. **1.6 feet**
    The shape of each storage tank is a cylinder, and the volume of a cylinder is given by $\pi r^2 h$, where r is the radius and h is the height. Jean-Luc wants a storage tank that is 5 feet 3 inches high. Convert this height to feet: 5 3/12 feet, or 5.25 feet. With $h = 5.25$, we can write the volume formula as $v = 5.25\pi r^2$. To write a function for r in terms of v, solve this equation for r.

    $$\frac{v}{5.25\,\pi} = r^2 \qquad\qquad \text{Divide both sides by 5.25π.}$$

    $$+\sqrt{\frac{v}{5.25\pi}} = r \qquad\qquad \text{Take the square root of both sides.}$$

    In this situation, r is a measurement of the radius, so it must be positive.

    Instead of volume in cubic feet, we are given the capacity of the water tank in gallons, 80 gallons. We must convert gallons to cubic feet to substitute for v in our formula.

    $$80 \text{ gallons} \div 7.48 \ \frac{\text{gallons}}{\text{cubic foot}} \approx 10.7 \text{ cubic feet}$$

    $$r = \sqrt{\frac{10.7}{5.25\pi}} \qquad\qquad \text{Substitute 10.7 for } v \text{ in the radius function.}$$

    $$r \approx 0.8 \qquad\qquad \text{Use your calculator to evaluate } \sqrt{\frac{10.7}{5.25\pi}}.$$

    The diameter is twice the radius: $2 \cdot 0.8 = 1.6$. Jean-Luc chose a storage tank that is 1.6 feet in diameter.

REFLECT

Congratulations on completing Chapter 3!
Here's what we just covered.
Rate your confidence in your ability to:

- Solve single-variable rational and radical equations and inequalities

  ① ② ③ ④ ⑤

- Write and solve rational and radical equations, inequalities, and functions to describe real-life situations and solve problems

  ① ② ③ ④ ⑤

- Graph rational and radical functions

  ① ② ③ ④ ⑤

- Use systems of equations to solve single-variable rational and radical equations

  ① ② ③ ④ ⑤

If you rated any of these topics lower than you'd like, consider reviewing the corresponding lesson before moving on, especially if you found yourself unable to correctly answer one of the related end-of-chapter questions.

Access your online student tools for a handy, printable list of Key Points for this chapter. These can be helpful for retaining what you've learned as you continue to explore these topics.

Chapter 4
Trigonometric Functions

GOALS By the end of this chapter, you will be able to

- Use the relationship between radius, radian measure, and arc length to solve for any one of these values for a given circle and central angle

- Convert between degrees and radians for any angle measure

- Find the value of the sine, cosine, tangent, cosecant, secant, or cotangent, where it exists, for any real number

- Graph sine, cosine, and tangent functions using information about the period, midline, amplitude, horizontal shift, and any reflection, and write an equation to describe a given sine, cosine, or tangent function

- Prove and apply trigonometric function identities

- Write and use sine and cosine functions to represent real-life situations and solve problems

Lesson 4.1
Trigonometric Ratios and Triangles

REVIEW

Trigonometric functions can be expressed as ratios of side lengths in a right triangle. For an acute angle θ:

Sine: $\sin\theta = \dfrac{\text{opposite}}{\text{hypotenuse}}$

Cosine: $\cos\theta = \dfrac{\text{adjacent}}{\text{hypotenuse}}$

Tangent: $\tan\theta = \dfrac{\text{opposite}}{\text{adjacent}}$

The **Pythagorean theorem** states that the sum of the squares of the leg lengths of a right triangle is equal to the square of its hypotenuse: $a^2 + b^2 = c^2$.

Let's review the basic trigonometric relationships as ratios of side lengths in right triangle *ABC*, shown below.

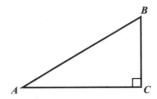

The hypotenuse is side *AB*, so $\sin \angle A = \dfrac{BC}{AB}$, $\cos \angle A = \dfrac{AC}{AB}$, and $\tan \angle A = \dfrac{BC}{AC}$.

Looking at the other acute angle, $\sin \angle B = \dfrac{AC}{AB}$, $\cos \angle B = \dfrac{BC}{AB}$, and $\tan \angle B = \dfrac{AC}{BC}$.

Notice that $\sin \angle A = \cos \angle B$ and $\cos \angle A = \sin \angle B$. The sine of an acute angle is always equal to the cosine of its complementary angle.

These trigonometric ratios are useful for solving problems involving right triangles, as you learned in Geometry.

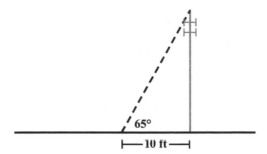

Paul wants to know how tall a certain cell phone tower is. Paul walks 10 feet from the cell phone tower, lies on the ground, and measures an angle of 65° from the ground to the top of the tower. Approximately how tall is the cell phone tower from the ground to its top?

We want to find the height of the cell phone tower, which is the leg of the right triangle that is opposite the 65° angle. The given length of 10 feet is the leg adjacent to the 65° angle. The trigonometric function that relates the opposite to the adjacent side is the tangent function. Be sure to put the opposite side length in the numerator and the adjacent side length in the denominator to represent the tangent.

$\tan 65° = h/10$	(h represents the height of the cell phone tower, in feet)
$10 \tan 65° = h$	Multiply both sides by 10.
$10(2.1445...) = h$	Use your calculator to find tan 65°.
$21 \approx h$	Multiply and round the product to the nearest whole number.

The cell phone tower is about 21 feet tall.

There are six basic trigonometric functions, including sine, cosine, and tangent, and their reciprocals, cosecant, secant, and cotangent.

Cosecant: $\csc \theta = \dfrac{1}{\sin \theta} = \dfrac{\text{hypotenuse}}{\text{opposite}}$

Secant: $\sec \theta = \dfrac{1}{\cos \theta} = \dfrac{\text{hypotenuse}}{\text{adjacent}}$

Cotangent: $\cot \theta = \dfrac{1}{\tan \theta} = \dfrac{\text{adjacent}}{\text{opposite}}$

EXAMPLE 2

In the right triangle shown below, csc θ is 13/12. What is cos θ?

The cosecant is the reciprocal of the sine of an angle, and vice versa. We know that csc θ = 13/12, so sin θ = 12/13. However, the question asks for the cosine of θ.

While you can work with 12 and 13 as representative numbers, keep in mind that these may not be the actual side lengths. Trigonometric functions are only ratios, so it could be that the opposite side is 24 and the hypotenuse is 26, or any other pair of real numbers with the same ratio.

Let's use the given information to label the sides of the triangle. Because csc θ = 13/12, we know that the ratio of the hypotenuse to the opposite side is 13 to 12. Label the hypotenuse as 13 and the side opposite θ as 12.

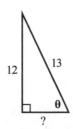

The length of the leg adjacent to θ is unknown, but we can solve for it using the Pythagorean theorem.

$a^2 + 12^2 = 13^2$
$a^2 + 144 = 169$ Find both squares.
$a^2 = 25$ Subtract 144 from both sides.
$a = 5$ Solve for the positive value of a. Side
                                              lengths are always positive.

The missing side length is 5 units. Use this to find cos θ.

$$\cos \theta = \frac{\text{adjacent}}{\text{hypotenuse}} = 5/13$$

Lesson 4.2
Radians

REVIEW

Angles can be measured from 0º to 360º. Those that are less than 90º are called **acute**, while those that are between 90º and 180º are **obtuse**. **Right** angles are exactly 90º, and **straight** angles are exactly 180º.

Triangles can be defined either by their angles (a **right triangle** has a right angle) or sides (an **isosceles triangle** has two equal sides).

A **radius** of a circle is any line segment connecting a point on the circle to the center of the circle. The radius of a circle is also defined as the length of any radius of the circle.

The **diameter** of a circle is a line segment passing through the center of the circle, with both endpoints on the circle. The length of the diameter is twice the radius.

The **circumference** of a circle is the distance around the circle and is equal to $2\pi r$, where r is the length of the radius of the circle. An **arc** is a portion of that circumference.

A **30-60-90 triangle** is a right triangle with angle measures of 30°, 60°, and 90°, and has side lengths in the ratio $1:\sqrt{3}:2$, respectively opposite each of those angles.

Degrees and radians are two different units used to express angle measurements. Radians are defined in relationship to a circle, as measuring a central angle of the circle. A **central angle** is an angle formed by two radii of the circle. Its vertex is the center of the circle.

> The radian measure of a central angle of a circle is the ratio of the length of the circular arc the angle subtends to the radius of the circle.

The circular arc **subtended** by the central angle is the portion of the circle between the endpoints of the radii that form the angle. In the diagram below, the central angle is labeled as θ, a radius is labeled as r, and the subtended arc is labeled as s.

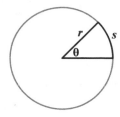

All radii of a circle are congruent, so the unlabeled radius forming the other side of the central angle is the same length, *r*.

So, the measure of θ in radians is *s/r*, where *s* is the arc length and *r* is the radius.

A **minor arc** is an arc subtended by a central angle of less than 180°, and a **major arc** is an arc subtended by a central angle of more than 180°. For any given two points on a circle, the minor arc and the major arc connecting them make up the entire circle. The minor arc is the shorter path between them, and the major arc is the longer path.

When two points on a circle are endpoints of the same diameter, then the two arcs connecting them are each exactly 180° and neither is minor nor major.

EXAMPLE

In the diagram below, circle *P* has a diameter of 12 units. If minor arc *QR* has a measure of 4π, what is the measure of ∠*QPR* in radians?

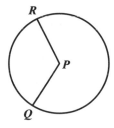

Minor arc *QR* is the shorter arc connecting points *Q* and *R*. We know from the definition of radian measure that the measure of angle *QPR* in radians equals *s/r*. The arc length of arc *QR* is 4π. The diameter of circle *P* is 12, so its radius is 6.

m∠*QPR* = *s/r* — Use the definition of radian measure to write an equation.

m∠*QPR* = 4π/6 — Substitute 4π for *s* and 6 for *r*.

m∠*QPR* = 2π/3 — Simplify the fraction.

The measure of angle *QPR* is 2π/3 radians.

The expression 2π/3 is the same as 2/3 π. The π can be written as part of the fraction, in the numerator, or can follow the fraction.

RADIANS AND DEGREES

A complete revolution of a circle is 360 degrees (360°). According to the definition of radian measure, the radian measure of a complete revolution of a circle is equal to the

Divide both sides of the ratio 360:2π by 2 to get the simplified ratio 180:π.

circumference of the circle divided by its radius. The circumference of a circle is given by the expression $2\pi r$, so the radian measure of a full-circle central angle is $\frac{2\pi r}{r}$, which simplifies to 2π. So, 360° is equivalent to 2π radians.

> $180° = \pi$ radians

EXAMPLE 4

25/11 π radians is approximately equal to how many degrees?

To convert from radians to degrees, we must multiply by 180/π.

$$\frac{\frac{25}{11}\pi \text{ radians}}{1} \times \frac{180 \text{ degrees}}{\pi \text{ radians}}$$

Converting between radians and degrees is just like switching between units of measurement like inches and feet. Remember to carefully label units as you convert, and cancel out units that overlap. In this example, radians will cancel out, leaving degrees.

$$\frac{25\cancel{\pi}}{11} \cdot \frac{180}{\cancel{\pi}}$$

Cancel out the common factor of π in the numerator and denominator.

Here is how you may see radians on the SAT.

In the figure to the right, circle O has a radius of 4, and angle XOY measures $\frac{7}{12}\pi$ radians. What is the measure of minor arc XY?

A) $\frac{7}{12}\pi$

B) $\frac{7}{3}\pi$

C) 7π

D) 12π

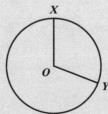

$$\frac{25 \cdot 180}{11}$$
Multiply to combine fractions.

Notice that the angle measure is greater than 360°, or 2π. It is possible, in trigonometry, to have angles that measure more than a full revolution of a circle.

$408.\overline{09}$
Calculate fraction value.

25/11 π is approximately equal to 409°.

EXAMPLE 5

What is the measure, in radians, of an acute angle of an isosceles right triangle?

An isosceles right triangle is also known as a 45-45-90 triangle, because its angles measure 45°, 45°, and 90°. Each acute angle measures 45°, so we must convert 45° to radians.

45 degrees × $\dfrac{\pi \text{ radians}}{180 \text{ degrees}}$

45π/180
Multiply.

π/4
180 is the same as 4 · 45, so we can cancel out the common factors of 45.

This time π is on top and 180 is on the bottom, because we are "canceling out" degrees in the unit labels, converting to radians.

An acute angle of an isosceles right triangle, or a 45° angle, has a measure of π/4 radians.

RADIANS AND UNIT CIRCLES

Although radians measure central angles of circles of any radius (such as in Example 3), mathematicians tend to use the **unit circle**—a circle of radius 1 unit—for work involving radians. Why is that? Well, take a look at the relationship of radians to arc length and radius, θ = s/r. When the radius, r, equals 1, the equation becomes simply θ = s.

> The radian measure of an angle is equal to the length of the arc on the unit circle subtended by the angle.

EXAMPLE 6

Shown below is a sector of a unit circle. What is the length of arc *EF*?

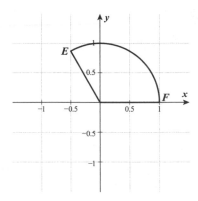

To find the length of the arc, we have to find the measure of its central angle, in radians. We know that it is greater than $\pi/2$, or 90°, because the section in the first quadrant, between the *x*-axis and *y*-axis, is a right angle. The measure of the additional portion to the left of the *y*-axis is what we must find.

Notice that point *E* has an *x*-value of –1/2, which means that it is exactly 1/2 unit away from the *y*-axis. When we draw this direct distance as a horizontal line, we form a triangle with the angle side and the *y*-axis.

We know that the measure of angle *XOY* in radians equals *s/r*. The angle measure is 7/12 π radians, and the radius is 4.

$$m\angle XOY = s/r$$

$$\frac{7}{12}\pi = \frac{s}{4}$$
Substitute 7/12 π for m∠*XOY* and 4 for *r*.

$$4 \cdot \frac{7}{12}\pi = s$$
Multiply both sides by 4.

$$\frac{7}{3}\pi = s$$
Cancel out the common factor of 4 in the numerator and denominator.

Minor arc *XY* has a measure of 7/3 π. The correct answer is (B).

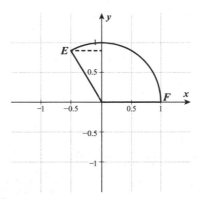

This is a right triangle, because the horizontal dashed line and the vertical *y*-axis are perpendicular. The length of the diagonal side of the triangle is 1 unit, because it is a radius of a unit circle. The short side is 1/2 unit long, because this is the distance from point *E* to the *y*-axis.

Does this look familiar? When the hypotenuse of a right triangle is twice as long as its shorter leg, you know that it must be a 30-60-90 right triangle. In our triangle, the hypotenuse length, 1, is 2 times the leg with length 1/2, so the angle opposite that leg measures 30°.

Now we know that the angle of the sector is 90° + 30° = 120°. Convert this measure to radians.

$$120° \times \frac{\pi \text{ radians}}{180°} = 2/3 \, \pi \text{ radians}$$

The angle of the sector measures 2/3 π, so the length of the unit circle arc *EF* is also 2/3 π.

Lesson 4.3
Trigonometric Functions and Circles

When we looked at trigonometric functions in relationship to triangles, we could only define them for angles between 0 and 90 degrees, or 0 and $\pi/2$ radians.

> Using a unit circle in the coordinate plane allows us to evaluate trigonometric functions (where they exist) for all real numbers. We will express these real numbers in terms of radian measures of central angles of the unit circle, measured counterclockwise from the positive x-axis.

This means that one side of the angle is always fixed on the positive x-axis. This is the **initial side** of the angle. The **terminal side** is the other side forming the angle, and it can be rotated anywhere along the circle. Both the initial side and the terminal side have a length of 1, because they are both radii of the unit circle. Let's look at angles with measures of $\pi/4$, $\pi/2$, π, $5\pi/4$, and $9\pi/4$.

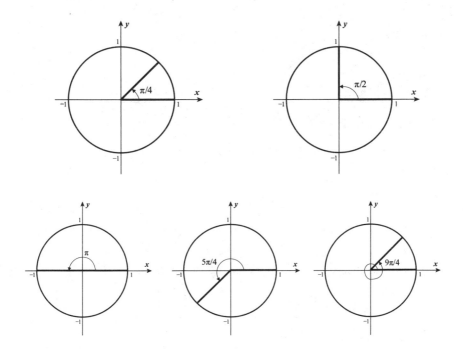

The initial side is always stationary. The terminal side rotates counterclockwise the given number of radians to form the angle. An angle of measure $\pi/2$ is a right angle, and an angle of measure π is a straight angle. An angle of measure 2π (not shown) is a full revolution of the circle, so any angle with a greater measure continues into a second revolution of the circle, as with $9\pi/4$. This angle looks like $\pi/4$ because it is **coterminal** with $\pi/4$, meaning that their terminal sides are the same. This is because $9\pi/4 = 2\pi + \pi/4$: a full revolution plus an additional rotation of $\pi/4$.

Angles defined this way can also have negative measures, which represent a clockwise rotation of the terminal side.

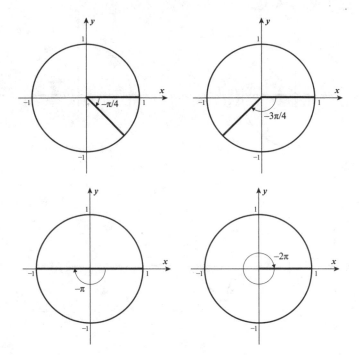

Notice that a −3π/4 angle is coterminal with a 5π/4 angle, and a −π angle is coterminal with a π angle.

> Coterminal angles have a difference of
> measure that is a multiple of 2π.

This holds true for each of the three coterminal angle pairs that we have looked at so far.

$\pi - (-\pi) = \pi + \pi = 2\pi$

$5\pi/4 - (-3\pi/4) = 5\pi/4 + 3\pi/4 = 8\pi/4 = 2\pi$

$9\pi/4 - \pi/4 = 8\pi/4 = 2\pi$

If we looked at the angle 5π, it would be coterminal with π. The difference between 5π and π is 4π, which is 2 times 2π. An angle of 5π makes two full revolutions of the unit circle and then rotates an additional π radians to form a straight angle.

For any angle measure, you can determine the trigonometric function value, if it exists, using the unit circle. The sine and cosine exist for all real numbers.

EXAMPLE 7

Using central angles formed by radii of a unit circle, find the values of sin $\pi/4$, sin $\pi/2$, sin $3\pi/4$, sin π, and sin $7\pi/6$.

Let's look at the angle $\pi/4$, adding a vertical line to complete a right triangle with the terminal side and part of the initial side.

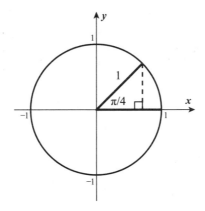

The sine value is $\dfrac{\text{opposite}}{\text{hypotenuse}}$. Any radius of a unit circle is 1 unit long, so the

hypotenuse of this right triangle is 1, which means the sine is equal to the opposite

(vertical) side. So, the sine of $\pi/4$ is the y-value of the point where the terminal side

of the angle reaches the unit circle. The degree equivalent of $\pi/4$ is 45°, so this is a

45-45-90 triangle, which means that the leg-to-hypotenuse ratio is 1: $\sqrt{2}$. Let's set

up a proportion with leg lengths in the numerators and hypotenuse lengths in the

denominators.

$y/1 = 1/\sqrt{2}$

The ratio of opposite leg length to radius 1 is equal to the ratio $1:\sqrt{2}$.

$y = 1/\sqrt{2}$

Rewrite $y/1$ as y.

$y = \sqrt{2}/2$

Multiply by $\sqrt{2}/\sqrt{2}$ to rationalize the denominator.

The height of the endpoint of the terminal side of the angle is $\sqrt{2}/2$, so the sine of $\pi/4$ is $\sqrt{2}/2$.

For any central angle, you can draw a vertical line to form a right triangle. The sine of the angle is equal to the y-value of the point where the terminal side meets the unit circle, because the hypotenuse of the right triangle formed is always 1.

So, sin θ ranges from 0 when θ = 0 to 1 when θ = π/2, as indicated by the right triangles of increasing height shown in the following figure.

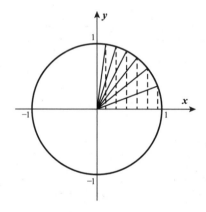

For a central angle of π/2, the leg "opposite" the central angle in the right triangle is actually the same length as the hypotenuse, 1. This is not literally a triangle at this point, because the adjacent side (the horizontal leg) has shrunk to 0, but the sine value still exists for π/2. So, sin π/2 = 1.

Here is the central angle $3\pi/4$. Notice that we still draw a vertical line from the x-axis to the end of the terminal side of the angle to form a right triangle. Now the sine is calculated using the acute angle of the triangle, even though we are finding the sine of $3\pi/4$.

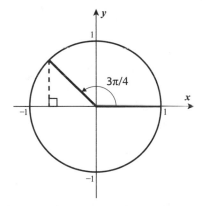

The sine of $3\pi/4$ is the sine value of the acute angle that is supplementary to $3\pi/4$, or $\pi/4$. Again, the sine is equal to the y-value of the point where the terminal side of the angle meets the unit circle, because this is the opposite side of the right triangle formed with the x-axis and terminal side. So, $\sin 3\pi/4 = \sqrt{2}/2$.

The sine values are decreasing in this quadrant, for angle measures from $\pi/2$ to π. The central angle π is a straight angle, or a straight line, as shown below.

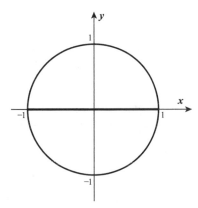

For a straight angle, the height of the opposite leg is 0, and the y-value of the endpoint of the terminal side is 0. So, $\sin \pi = 0$.

The angle $7\pi/6$ is shown below.

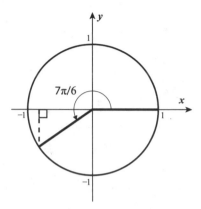

Here, the right triangle is below the x-axis, where y-values are negative. The sine of the central angle is still equal to the y-value of the point where the terminal side meets the unit circle, so in this case it is negative.

The angle $7\pi/6$ is a straight angle (π) plus an additional $\pi/6$ angle. The degree equivalent of $\pi/6$ is 30°, so the right triangle shown is a 30-60-90 triangle. The side opposite the $\pi/6$ angle has a length that is 1/2 the hypotenuse. The hypotenuse is 1 unit long, so the opposite leg is 1/2 unit long, and the y-value of the endpoint is −1/2. So, $\sin 7\pi/6 = -1/2$.

For all angles greater than π and less than 2π, the sine value is negative. From π to $3\pi/2$, the sine values decrease from 0 to −1, and from $3\pi/2$ to 2π, the sine values increase from −1 to 0.

Remember that coterminal angles have the same terminal side, so they also have the same sine value. For example, $\sin (-5\pi/6) = \sin 7\pi/6 = -1/2$.

> In fact, because coterminal angles share the same terminal side, they have the same function value for each of the trigonometric functions.

Lesson 4.4
Graphing Trigonometric Functions

If we plot sine values using a standard (x, y) coordinate plane, with angle measure, in radians, on the x-axis and sine values on the y-axis, we can see the graph of the basic sine function, $y = \sin x$.

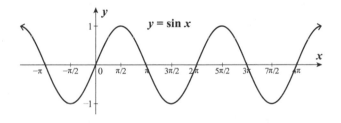

All of the sine values found in Example 7 are shown in the graph: $(0, 0)$, $(\pi/4, \sqrt{2}/2)$, $(\pi/2, 1)$, $(3\pi/4, \sqrt{2}/2)$, $(\pi, 0)$, $(7\pi/6, -1/2)$, $(3\pi/2, -1)$, and $(2\pi, 0)$.

The function curve repeats the same wave pattern again and again, endlessly. This sine curve is a **periodic function** with a **period** of 2π, which means that the pattern repeats every 2π units on the x-axis. This is because a full revolution of the unit circle is 2π radians. Angle measures 2π apart are coterminal angles, with the same sine value.

Graph the basic cosine function, $y = \cos x$.

As when we were finding sine values, we can draw a vertical line from the point where the terminal side of a central angle meets the unit circle to the x-axis to create a right triangle.

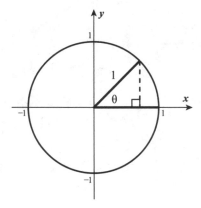

The cosine of an angle is $\dfrac{\text{adjacent}}{\text{hypotenuse}}$, and the hypotenuse, a radius of the unit circle,

is equal to 1, so the cosine of the angle is the length of the adjacent side, the portion

of the initial angle side that is part of the triangle formed.

> The cosine of any central angle is equal to the x-value of the point where the terminal side meets the unit circle.

Imagine how the x-value of the point where the terminal side meets the unit circle changes for angles measuring 0 to $\pi/2$ radians. Some x-values in this range are represented by the horizontal legs of the triangles shown below.

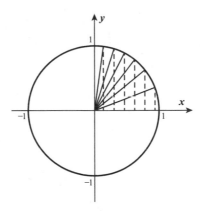

The cosine of 0 is 1, because the terminal side lies on the initial side, extending to 1 on the *x*-axis. The cosine value decreases from there until an angle of $\pi/2$, at which point the cosine value has shrunk to 0.

To the left of the *y*-axis, *x*-values are negative, so the cosine values of angles between $\pi/2$ and $3\pi/2$ are negative. From cos $\pi/2$ at 0, cosine values decrease to –1 for an angle of π. From π to $3\pi/2$, cosine values increase from –1 to 0.

In Quadrant IV, *x*-values are positive, so cosine values are positive. From $3\pi/2$ to 2π, the cosine increases from 0 to 1.

Also, recall that the cosine of an angle is equal to the sine of its complementary angle, as we saw with triangle *ABC* at the beginning of this chapter. To put it another way, the cosine of an angle θ is equal to the sine of $(\pi/2 - \theta)$, and this is true for all angles (not just angles between 0 and $\pi/2$). So, the cosine function graph should have the same basic shape of the sine function graph but passing through the values we've found, (0, 1), ($\pi/2$, 0), (π, –1), ($3\pi/2$, 0), and (2π, 1).

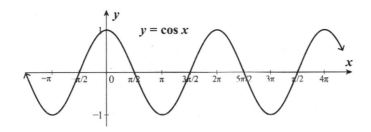

The tangent function is different, because its ratio does not include the hypotenuse. Tangent = $\dfrac{\text{opposite}}{\text{adjacent}}$. We can still draw a vertical line from the endpoint of the terminal side of the angle to the x-axis to form a right triangle, which we can use to calculate the tangent value.

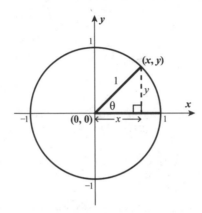

The opposite side is equal to the y-value, and the adjacent side is equal to the x-value of the point where the terminal side meets the unit circle. The terminal side of the angle begins at the origin, (0, 0). So, the tangent of this angle is equal to y/x, or the change in y-values (from 0 to y) divided by the change in x-values (from 0 to x). In other words, the tangent of a central angle is equal to the slope of its terminal side.

When θ = 0, tan θ = 0/1 = 0. When θ = π/4, the triangle formed is an isosceles right triangle, so x = y and tan π/4 therefore equals 1. When θ = π/2, tan θ = 1/0. This fraction is undefined, so the tangent of π/2 is undefined. In other words, tan π/2 does not exist. However, as θ increases and approaches π/2, the adjacent leg becomes very tiny, causing the tangent value to become very large, approaching ∞.

In Quadrant II, the opposite leg (y-value) is positive, but the adjacent leg (x-value) is negative. So, the tangent values of angles between π/2 and π are negative, ranging from close to −∞ to close to 0. The value of tan π is 0/1, or 0.

In Quadrant III, both the opposite leg (*y*-value) and the adjacent leg (*x*-value) are negative, as shown below, so the tangent value is positive. A negative number divided by a negative number produces a positive quotient.

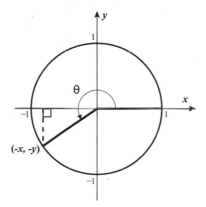

The tangent of $3\pi/2$ is $-1/0$, an undefined value, so tan $3\pi/2$ does not exist.

In Quadrant IV, the opposite leg (*y*-value) is negative, and the adjacent leg (*x*-value) is positive. So, the tangent values of angles between $3\pi/2$ and 2π are negative and increasing to 0, the value of tan 2π.

The graph of the basic tangent function, *y* = tan *x*, is shown below.

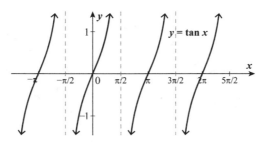

There are vertical asymptotes at $-\pi/2$, $\pi/2$, and $3\pi/2$, as indicated by the dashed lines. The tangent value does not exist for these angle measures.

Notice that the curve pattern for tangent repeats every π units.

The tangent function has a period of π units, whereas the sine and cosine functions have periods of 2π units.

CHANGES TO PERIOD, MIDLINE, AND AMPLITUDE OF GRAPHS

The graphs of the sine, cosine, and tangent functions that we have seen so far are the parent functions, the most basic versions of each trigonometric function. As with polynomial, rational, or radical functions, additional constant terms or coefficients affect the position and scale of the graph.

EXAMPLE 9

Graph the function $y = \sin 2x$.

We know that $\sin 0 = 0$, $\sin \pi/2 = 1$, $\sin \pi = 0$, $\sin 3\pi/2 = -1$, and $\sin 2\pi = 0$. We can use these to find points on the function $y = \sin 2x$.

When $x = 0$, $y = \sin (2 \cdot 0) = \sin 0 = 0$.

When $x = \pi/4$, $y = \sin (2 \cdot \pi/4) = \sin \pi/2 = 1$.

When $x = \pi/2$, $y = \sin (2 \cdot \pi/2) = \sin \pi = 0$.

When $x = 3\pi/4$, $y = \sin (2 \cdot 3\pi/4) = \sin 3\pi/2 = -1$.

When $x = \pi$, $y = \sin (2 \cdot \pi) = \sin 2\pi = 0$.

The graph of $y = \sin 2x$ passes through all the same y-values as the graph of $y = \sin x$ but gets to them twice as quickly. The graphs of $y = \sin 2x$ and $y = \sin x$ are shown below.

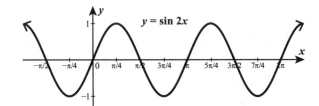

Looks can be deceiving! Always use the values along the x-axis, rather than the overall appearance of the graph, to assess the function's period. In these graphs, we have "zoomed in" (horizontally only) to better compare the two sine functions, so this graph of $y = \sin x$ appears more spread out than the one at the beginning of this lesson, but it actually has the same period.

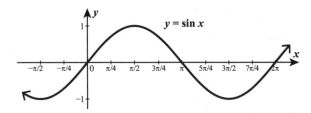

The graph of $y = \sin 2x$ completes a full cycle in half the distance of the parent sine curve, $y = \sin x$. The period of $y = \sin 2x$ is π, which is $1/2$ the period of $y = \sin x$.

The **frequency** of a trigonometric function is the number of cycles it completes in a given interval, so the period and frequency of a trigonometric function are inversely related. As the period increases, the frequency decreases. As the period decreases, the frequency increases, as in Example 9. For a domain interval of 2π radians, $y = \sin x$ has a frequency of 1 (one complete wave cycle), but $y = \sin 2x$ has a frequency of 2 (two complete wave cycles). Halving the period doubled the frequency.

For all of the trigonometric functions we have graphed so far, the midline has been the x-axis. The **midline** is the horizontal line around which the periodic function oscillates and is halfway between the maximum and minimum for sine and cosine functions. The midline of the tangent function is the horizontal line that includes the points of inflection of the graph. A **point of inflection** is a point where a curve changes its direction, from concave to convex or from convex to concave.

Some trigonometric functions have midlines that are above or below the x-axis.

Contrast this with a coefficient of x that is less than 1. The period of $y = \sin 1/2\ x$, or $y = \sin x/2$, is 2 times the period of $y = \sin x$. It stretches the full cycle across 4π units.

The hertz (Hz) is the unit of frequency defined as one cycle per second. In science, the frequencies of sound waves, light waves, and radio waves are usually measured in reference to a time unit, such as a second. Because there are such a huge number of waves per second, the frequencies of these waves are defined in kilohertz (10^3 Hz), gigahertz (10^9 Hz), or terahertz (10^{12} Hz).

EXAMPLE 10

What equation could describe the function graphed below?

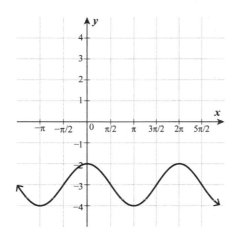

The maximum value of the function is −2, and the minimum value is −4, so the midline is the horizontal line at the average of these y-values.

$$\frac{-2+\left(-4\right)}{2} = -6/2 = -3$$

This function has a midline of y = −3.

Looking at the midline can immediately tell you the shift of the graph. The midline y = −3 means that the parent cosine graph was shifted 3 units down, and −3 is the constant to add to y = cos x when writing the function equation.

The function is at its maximum when it crosses the y-axis, which is also the case for y = cos x. Compare this graph to the one of y = cos x in Example 8. The graphs have the same shape and the same period, but the one shown above is the graph of y = cos x shifted 3 units directly down. For any given x-value, this graph has a y-value 3 units less than y = cos x. Where cos x = 1, this graph is at −2; where cos x = 0, this graph is at −3; and where cos x = −1, this graph is at −4.

The function graphed here is y = cos x − 3. As with other kinds of functions, adding a constant to the original function value vertically shifts the function graph that number of units: up if the constant is positive, down if the constant is negative.

As discussed in Lesson 1.4, f(x − c) represents f(x) shifted c units horizontally. Watch the sign! When c is negative, as in f(x − [− c]), it becomes f(x + c) and shifts to the left.

We just saw how a trigonometric function graph can be shifted vertically. Trigonometric function graphs can also be shifted horizontally, by adding a constant to the x term within the function. In other words, cos (x + 2) represents a cosine function shifted 2 units horizontally, to the left.

EXAMPLE

Graph the function y = tan (x − π/2).

Look at the graph of y = tan x to determine some values along the curve y = tan (x − π/2). When x = 0, the expression tan (x − π/2) becomes tan (−π/2), which is undefined. When x = π/2, the expression tan (x − π/2) becomes tan 0, which equals 0. For each x-value, the function y = tan (x − π/2) is the same as the function y = tan x exactly π/2 units to the left of there. So, the function y = tan (x − π/2) is the function y = tan x shifted π/2 units to the right.

The parentheses tell us that π/2 is subtracted from x, not from tan x. If you were asked to graph the function y = tan x − π/2, it would instead be the graph of y = tan x shifted vertically π/2 units down.

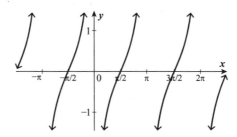

This graph is also the graph of $y = \tan (x + \pi/2)$, because it is the graph of $y = \tan x$ shifted $\pi/2$ units to the left. The reason $y = \tan (x - \pi/2)$ and $y = \tan (x + \pi/2)$ look identical is because they are exactly one full tangent period apart. The difference $(x + \pi/2) - (x - \pi/2)$ is equal to π, the period of the basic tangent function.

In addition to shifting horizontally and vertically and expanding and contracting the cycles (changing the period), we can also adjust the amplitude of a trigonometric function. The **amplitude** is the distance between the midline and the maximum or minimum for sine and cosine functions. Tangent functions have no maximum or minimum, so they do not have amplitudes, but their scale can be similarly adjusted.

You can also think of the amplitude as the height of the wave of the function, from the midline. Just remember that distance is always positive, so the amplitude is always positive, even when you measure the dip of a wave below the midline.

EXAMPLE

Graph the function $y = 4 \sin x$.

First, remember what the function $y = \sin x$ looks like.

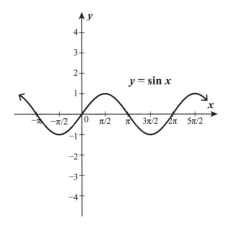

For each value of x, $4 \sin x$ will be 4 times the value of $\sin x$. For example, when $x = \pi/2$, $\sin x = 1$, so $4 \sin x = 4$. This is where the sine curves reach their maximum. When $x = \pi/6$, $\sin x = 1/2$, so $4 \sin x = 2$. When $x = \pi$, $\sin x = 0$, so $4 \sin x$ also equals 0. The graph of $y = 4 \sin x$ is shown below.

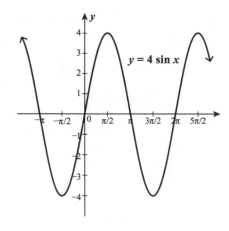

The amplitude of $y = \sin x$ is 1, because this is the distance between the midline value (0) and either the maximum (1) or the minimum (−1). The graph of $y = 4 \sin x$ also has a midline of $y = 0$ (the x-axis), but its maximum is 4 and its minimum is −4. The amplitude of $y = 4 \sin x$ is 4. The amplitude of the parent graph was multiplied by the coefficient of the trigonometric function value.

EXAMPLE 13

Graph the function $y = -1/2 \tan x$.

First, let's review what $y = \tan x$ looks like.

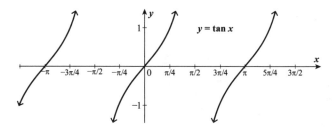

We can multiply various values of tan x by $-1/2$ to find points along the graph of the function $y = -1/2 \tan x$.

When $x = -\pi$, $\tan x = 0$. So, $-1/2 \tan (-\pi) = 0$.

When $x = -3\pi/4$, $\tan x = 1$. So, $-1/2 \tan (-3\pi/4) = -1/2$.

When $x = -\pi/2$, $\tan x$ is undefined. So, $-1/2 \tan (-\pi/2)$ is also undefined.

When $x = -\pi/4$, $\tan x = -1$. So, $-1/2 \tan (-\pi/4) = 1/2$.

When $x = 0$, $\tan x = 0$. So, $-1/2 \tan 0 = 0$.

When $x = \pi/4$, $\tan x = 1$. So, $-1/2 \tan \pi/4 = -1/2$.

Here is the graph of $y = -1/2 \tan x$.

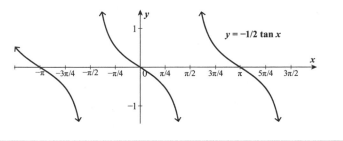

The coefficient of $1/2$ pulled all points closer to the midline, so the curves are more squat, although the period of the cycle is unchanged. The negative sign in $y = -1/2 \tan x$ changed the sign of the function value of $y = \tan x$ for each x-value, from positive to negative and from negative to positive. In other words, the negative sign in the coefficient reflected the function graph across the x-axis, so each continuous curve is always decreasing in $y = -1/2 \tan x$, whereas $y = \tan x$ is always increasing over each continuous section.

What happens when there is a negative sign in front of x instead of in front of the entire trigonometric expression? As in other types of functions, $f(-x)$ reflects the function $f(x)$ graph across the y-axis, because the function value for any given x-value is that of its opposite in the original function.

Try graphing $y = 1/2 \tan (-x)$ on a graphing calculator. It looks exactly the same as $y = -1/2 \tan x$. This is because the graph of $y = 1/2 \tan x$ is symmetric with respect to the origin. Reflecting it across the y-axis produces the same result as reflecting it across the x-axis.

However, reflections are not always across the x-axis or y-axis.

Graphs of the form $y = a \tan x$, where a is some real number, are symmetric with respect to the origin. However, once translated horizontally or vertically, the graph may not be symmetric with respect to the origin.

Graph the function $y = -\cos x + 1$.

The function has a midline of 1. Let's find some values along the curve.

Another way to approach this is to graph the function $y = -\cos x$, which is a reflection of $y = \cos x$ across the x-axis, and then translate it 1 unit upward.

When $x = 0$, $y = -\cos 0 + 1 = -(1) + 1 = 0$.

When $x = \pi/2$, $y = -\cos \pi/2 + 1 = -0 + 1 = 1$.

When $x = \pi$, $y = -\cos \pi + 1 = -(-1) + 1 = 2$.

When $x = 3\pi/2$, $y = -\cos 3\pi/2 + 1 = -0 + 1 = 1$.

When $x = 2\pi$, $y = -\cos 2\pi + 1 = -(1) + 1 = 0$.

Here is the graph of $y = -\cos x + 1$.

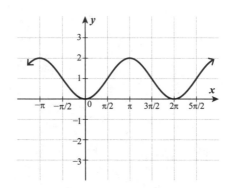

This is the graph of $y = \cos x + 1$ (the parent cosine graph shifted 1 unit up) reflected across the line $y = 1$. So, a negative sign in front of the trigonometric expression indicates a reflection of the function across the midline of the function. In $y = -1/2$ tan x in Example 13, the midline was the x-axis, so the graph was reflected across the x-axis.

As we mentioned in Example 8, $\cos \theta = \sin (\pi/2 - \theta)$, which can be written as $\sin -(\theta - \pi/2)$. So, the basic cosine function graph is the basic sine function graph shifted $\pi/2$ units to the right and then reflected across the line $y = \pi/2$.

Similarly, the graph of $y = \cos -(x - \pi)$ would be the parent graph shifted π units to the right and then reflected across the vertical line $x = \pi$. Or, you could rewrite it as $y = \cos (-x + \pi)$, which represents the parent cosine graph shifted π units to the left to form $f(x) = \cos x + \pi)$ and then reflected across the y-axis to form $f(-x)$, or $\cos (-x + \pi)$.

We can now summarize what we have learned about changes to the graphs of basic trigonometric functions.

A sine function of the form $y = a \sin b(x - c) + d$ or a cosine function of the form $y = a \cos b(x - c) + d$ is the respective parent trigonometric function adjusted as follows:

- The amplitude is equal to $|a|$. When $|a|$ is greater than 1, the waves are taller, reaching up and down $|a|$ units from the midline. When $|a|$ is less than 1, the waves are shorter than 1 unit.

- The graph is shifted vertically d units: up if d is positive, down if d is negative. The midline of the function is the line $y = d$.

- The graph is shifted horizontally c units: to the right if c is positive (when a constant is subtracted from x), to the left if c is negative (when a constant is added to x).

- The period of the graph is multiplied by $|1/b|$. So, if $|b|$ is greater than 1, the graph oscillates more quickly and looks more scrunched up, and if $|b|$ is less than 1, the graph oscillates more slowly and looks more spread out, left to right.

> The period of a sine or cosine function is $2\pi/b$, and the period of a tangent function is π/b.

- If a is negative, the graph is reflected across the midline, $y = d$ (the x-axis when $d = 0$). If b is negative, the graph is reflected across the line $x = c$ (the y-axis when $c = 0$).

All of the above also apply to a tangent function of the form $y = a \tan b(x - c) + d$ in relation to its parent function, except that a tangent function does not have an amplitude. Its curves, however, stretch vertically according to the value of a.

You will need to look at various attributes of a trigonometric function at the same time, rather than in isolation as we have done so far in this lesson.

EXAMPLE 15

Write an equation to describe the function graphed below.

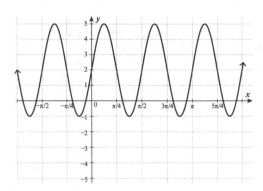

This graph shows a maximum of 5 and a minimum of −1. The average of these is

$\dfrac{5+(-1)}{2}$ = 2, so the midline for this graph is y = 2. The difference between the midline

value, 2, and the maximum, 5, or the minimum, −1, is 3. So, the amplitude is 3.

Another way to write an equation describing this graph is as a cosine function shifted horizontally. But, writing this as a sine function without any horizontal shift is simpler.

The graph begins at its midline, 2, when x = 0. The basic sine curve also starts at its midline when crossing the y-axis, so we'll write the equation of a sine function with an amplitude of 3 and translated 2 units up. However, the period is much shorter than 2π. This graph completes a full cycle in $\pi/2$ units, which is 1/4 of 2π. So, b in $y = a \sin b(x - c) + d$ is equal to 4.

A sine function with an amplitude, a, of 3, a vertical shift, d, of 2, and a period of $\pi/2$ (b = 4) is described by the equation y = 3 sin 4x + 2. This represents the given graph.

This equation can also be written as y = 2 + 3 sin 4x. By the commutative property of addition, the sum of the trigonometric expression and the constant d may be written in either order.

Graph the function $y = \cos(x + \pi/2)$. Then, write two different equations that can also be used to describe the same graph.

The constant $\pi/2$ is added to x, so the graph of $y = \cos(x + \pi/2)$ is the graph of $y = \cos x$ shifted $\pi/2$ units to the left. There are no other additional constants or coefficients. The graph of $y = \cos(x + \pi/2)$ is shown below.

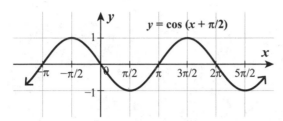

We know that this curve has a period of 2π, so a cosine function that is shifted 2π in either direction from $y = \cos(x + \pi/2)$ will look identical to this graph.

$$\cos(x + \pi/2 - 2\pi) = \cos(x - 3\pi/2)$$

Try graphing $y = \cos(x - 3\pi/2)$ to check. It looks exactly like the graph of $y = \cos(x + \pi/2)$, so $y = \cos(x - 3\pi/2)$ also describes this graph.

Notice, too, that this graph passes through the origin, which is the case with the parent sine graph. Compare this graph to the graph of $y = \sin x$. This graph is the reflection of $y = \sin x$ across the x-axis, with the same period, amplitude, and midline. So, the equation $y = -\sin x$ also describes this curve.

The functions
$y = \cos(x + 5\pi/2)$,
$y = \cos(x + 9\pi/2)$,
$y = \cos(x - 7\pi/2)$,
and all others shifted some multiple of 2π units horizontally also describe this graph.

We have found two additional equations to describe the curve, but there are many more. For example, this graph is also the reflection of $y = \sin x$ across the y-axis, so the equation $y = \sin(-x)$ also describes it.

This graph is also that of $y = \sin x$ shifted π units to the left or right, so the equations $y = \sin(x + \pi)$ and $y = \sin(x - \pi)$ each describe this curve as well. In fact, because the cycle repeats endlessly, there are an infinite number of equations that can accurately describe this curve. The same will be true of any given sine or cosine graph, though you'll generally want to work with the simplest answer.

Again, any function that is $y = -\sin x$ or $y = \sin(-x)$ shifted some multiple of 2π units is also represented by this graph. For example, the equations
$y = -\sin(x + 2\pi)$ and
$y = \sin(-x - 8\pi)$ each describe this graph.

Lesson 4.5
Trigonometric Function Identities

As with polynomials, there are identities that relate the various trigonometric functions to one another. We already described several of them earlier in this chapter, such as $\csc \theta = \dfrac{1}{\sin \theta}$, $\sec \theta = \dfrac{1}{\cos \theta}$, and $\cot \theta = \dfrac{1}{\tan \theta}$. Each of these represents an inverse relationship, so we can rewrite them as $\sin \theta = \dfrac{1}{\csc \theta}$, $\cos \theta = \dfrac{1}{\sec \theta}$, and $\tan \theta = \dfrac{1}{\cot \theta}$.

ACT Q

Here is how you may see trigonometric function graphs on the ACT.

Which of the following equations describes the equation graphed below?

A. $y = \sin x/2$

B. $y = -1 + \sin x/2$

C. $y = -1 + 2 \sin x/2$

D. $y = 2 \sin x$

E. $y = -1 + 2 \sin x$

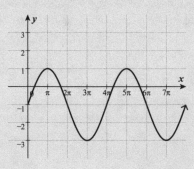

Write an identity for tan θ in terms of two other trigonometric functions.

Tangent is defined as $\frac{\text{opposite}}{\text{adjacent}}$. Another trigonometric function that involves the leg opposite the given angle is sine, which is $\frac{\text{opposite}}{\text{hypotenuse}}$. The cosine is defined as $\frac{\text{adjacent}}{\text{hypotenuse}}$. Let's divide sin θ by cos θ.

$\frac{\text{opposite}}{\text{hypotenuse}} \div \frac{\text{adjacent}}{\text{hypotenuse}}$	Write each trigonometric function as its ratio.
$\frac{\text{opposite}}{\text{hypotenuse}} \times \frac{\text{hypotenuse}}{\text{adjacent}}$	Rewrite the division as multiplication by the reciprocal of the divisor.
$\frac{\text{opposite}}{\cancel{\text{hypotenuse}}} \times \frac{\cancel{\text{hypotenuse}}}{\text{adjacent}}$	Cancel out the common factor in the numerator and denominator.
$\frac{\text{opposite}}{\text{adjacent}}$	Multiply.
tan θ	Use the definition of tangent.

Another way to write an identity for tan θ is $\tan\theta = \frac{\sec\theta}{\csc\theta}$. You can prove this using the same steps we used to prove that $\tan\theta = \frac{\sin\theta}{\cos\theta}$.

Another way to write sin θ ÷ cos θ is as the fraction $\frac{\sin\theta}{\cos\theta}$. So, our identity can be written as $\tan\theta = \frac{\sin\theta}{\cos\theta}$.

Another identity that we have already explored within this chapter is cos θ = sin (π/2 – θ). Likewise, sin θ = cos (π/2 – θ). Similar identities exist for the other trigonometric functions as well.

An easy way to remember these identities is to reference a right triangle, such as triangle *ABC* at the beginning of this chapter. The cosine of an angle is equal to the sine of its complementary angle. The sine of an angle is equal to the cosine of its complementary angle. These identities are just extended to all possible value of θ.

Here is a list of some useful trigonometric function identities.

Trigonometric Function Identities

$\sin \theta = \dfrac{1}{\csc \theta}$	$\sin \theta = \cos (\dfrac{\pi}{2} - \theta)$	
$\cos \theta = \dfrac{1}{\sec \theta}$	$\cos \theta = \sin (\dfrac{\pi}{2} - \theta)$	
$\tan \theta = \dfrac{1}{\cot \theta}$	$\tan \theta = \cot (\dfrac{\pi}{2} - \theta)$	$\tan \theta = \dfrac{\sin \theta}{\cos \theta}$
$\csc \theta = \dfrac{1}{\sin \theta}$	$\csc \theta = \sec (\dfrac{\pi}{2} - \theta)$	
$\sec \theta = \dfrac{1}{\cos \theta}$	$\sec \theta = \csc (\dfrac{\pi}{2} - \theta)$	
$\cot \theta = \dfrac{1}{\tan \theta}$	$\cot \theta = \tan (\dfrac{\pi}{2} - \theta)$	$\cot \theta = \dfrac{\cos \theta}{\sin \theta}$

As we saw in Example 16, these identities can also be extended to coterminal angles.

Any true equation that relates trigonometric functions is a trigonometric function identity.

ACT A

All of the answer choices are sine functions. The graph begins at the midline when $x = 0$, like $y = \sin x$, so this is a sine function shifted vertically downward.

The maximum of the graph shown is 1 and the minimum is −3. Let's find their average.

$$\frac{1+(-3)}{2} = -2/2 = -1$$

The midline is the line $y = -1$. The graph is shifted down 1 unit.

The amplitude is the difference between the midline value, −1, and the maximum, 1, or minimum, −3. So, the amplitude is 2.

Take a look at the period. This function has a period of 4π, so its period is twice that of the function $y = \sin x$. The coefficient of x must be 1/2. So, the function must be $y = -1 + 2 \sin (1/2 \, x)$, which can also be written as $y = -1 + 2 \sin x/2$. The correct answer is (C).

When squaring a trigonometric function, put the superscript immediately after the abbreviation. In other words, $\sin^2 x = (\sin x)^2$, or $\sin x \cdot \sin x$.

EXAMPLE

Fully simplify the expression $\dfrac{\tan x}{\sec x\ \sin^2 x}$ to find its equivalent trigonometric function, where the original expression exists.

Use the basic trigonometric function identities and the rules of fractions to simplify the given expression.

$$\dfrac{\dfrac{\sin x}{\cos x}}{\dfrac{1}{\cos x}\cdot \sin^2 x}$$

Replace $\tan x$ with $\dfrac{\sin x}{\cos x}$ and $\sec x$ with $\dfrac{1}{\cos x}$.

$$\dfrac{\dfrac{\sin x}{\cos x}}{\dfrac{\sin^2 x}{\cos x}}$$

Multiply the terms in the denominator.

$$\dfrac{\sin x}{\cos x}\times \dfrac{\cos x}{\sin^2 x}$$

Rewrite the division as multiplication by the reciprocal of the divisor.

$$\dfrac{\sin x\cdot \cancel{\cos x}}{\cancel{\cos x}\cdot \sin^2 x}$$

Multiply the fractions and cancel out the common factor of $\cos x$.

$$\dfrac{\cancel{\sin x}}{\sin x\cdot \cancel{\sin x}}$$

Rewrite $\sin^2 x$ as $\sin x \cdot \sin x$ and cancel out the common factor of $\sin x$.

This yields $\dfrac{1}{\sin x}$, and we can use a trigonometric identity to convert that to $\csc x$.

The expression $\dfrac{\tan x}{\sec x\ \sin^2 x}$, where it exists, is equivalent to $\csc x$.

Let's pause for a moment to consider the phrase "where it exists." Recall from work in previous chapters that if any part of the original expression would be undefined—whether that's from a denominator that's equal to zero, or because a trigonometric function like tan $\pi/2$ doesn't exist—then those values are not part of the solution set. That's why it's important that we include "where it exists," so as to acknowledge any constraints on an equivalence.

Another useful trigonometric function identity is a version of the familiar Pythagorean identity.

As we learned earlier, the sine value of any central angle centered at the origin is the y-value of the point where its terminal side intersects the unit circle, and its cosine value is the x-value of that point.

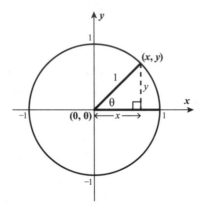

As shown in the diagram, the relationship between x, y, and 1 is a right triangle. By the Pythagorean theorem, $x^2 + y^2 = 1$. Even if the terminal side is in a different quadrant, where x and/or y is a negative number, the values still make the equation $x^2 + y^2 = 1$ true. Notice that this is the equation that describes the unit circle.

Here is how you may see the Pythagorean trigonometric identity on the ACT.

If $2x + \sin^2 \theta + \cos^2 \theta = 9$, then $x = ?$

    A. 1
    B. 2
    C. 3
    D. 4
    E. 5

Because $x = \cos\theta$ and $y = \sin\theta$, the equation can be rewritten as $\cos^2\theta + \sin^2\theta = 1$, or, using the commutative property of addition, as $\sin^2\theta + \cos^2\theta = 1$.

> The Pythagorean trigonometric identity states that $\sin^2\theta + \cos^2\theta = 1$ for any angle θ.

This identity allows you to solve for any trigonometric function value of an unknown angle if given the sine, cosine, or tangent of that angle and the quadrant in which it lies.

For an example of how trigonometric function identity questions appear on the ACT, please access the online Student Tools for this book.

EXAMPLE 19

For θ, an angle whose measure is between $\pi/2$ and π, $\cos\theta = -3/5$. What is $\sin\theta$? What is $\cot\theta$?

We are given the value of $\cos\theta$, $-3/5$. Substitute this into the Pythagorean trigonometric identity equation, $\sin^2\theta + \cos^2\theta = 1$.

$\sin^2\theta + (-3/5)^2 = 1$

$\sin^2\theta + 9/25 = 1$ Evaluate the square.

$\sin^2\theta = 16/25$ Subtract 9/25 from both sides, remembering that $1 = 25/25$.

$\sin\theta = \pm\sqrt{\dfrac{16}{25}}$ Take the square root of both sides.

$\sin\theta = \pm 4/5$ Evaluate the square root.

To determine whether $\sin\theta$ is positive or negative, look at what quadrant its terminal side lies in. The measure of θ is given to be between $\pi/2$ and π, so the terminal side is in Quadrant II, where x-values are negative and y-values are positive. The sine is the y-value of the point where the terminal side meets the unit circle, so here $\sin\theta$ is positive. The value of $\sin\theta$ is 4/5.

Use one of the identities for cotangent to solve for $\cot\theta$.

$\cot\theta = \dfrac{\cos\theta}{\sin\theta}$ Use the identity that uses cosine and sine, since we know these values.

$\cot\theta = \dfrac{-\dfrac{3}{5}}{\dfrac{4}{5}}$ Substitute the values we found for $\cos\theta$ and $\sin\theta$.

$\cot\theta = -3/5 \cdot 5/4$ Division by a fraction is multiplication by its reciprocal.

$\cot\theta = -3/4$ Multiply, canceling out the common factor of 5.

Solving the equation $\sin^2\theta + \cos^2\theta = 1$ for $\sin\theta$ gives you $\sin\theta = \pm\sqrt{1 - \cos^2\theta}$. Similarly, $\cos\theta = \pm\sqrt{1 - \sin^2\theta}$. You can use these equations as shortcuts when solving for $\sin\theta$ or $\cos\theta$. In this case, we have $\sin\theta = \pm\sqrt{1 - \left(-\dfrac{3}{5}\right)^2}$.

Remember that the value that was squared—in this case, $\sin\theta$—could be either positive or negative, so include the $\pm$ symbol when taking the square root.

Another way to determine the sign of sin θ is by looking at the graph of $y = \sin x$. When x is between $\pi/2$ and π, the value of $\sin x$ is always positive.

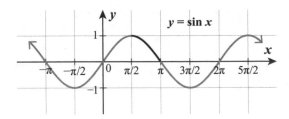

To solve for trigonometric function values of an unknown angle when given the tangent value of that angle and the quadrant of its terminal side, use the fact that tan $\theta = \dfrac{\sin \theta}{\cos \theta}$ in conjunction with the Pythagorean trigonometric identity.

Use the fact that $\sin^2 \theta + \cos^2 \theta = 1$ for any angle θ.

$2x + 1 = 9$	Substitute 1 for $\sin^2 \theta + \cos^2 \theta$.
$2x = 8$	Subtract 1 from both sides.
$x = 4$	Divide both sides by 2.

The correct answer is (D).

If tan θ = –8/15 and θ has a measure between 3π/2 and 2π, what is cos θ?

$\dfrac{\sin θ}{\cos θ}$ = –8/15

Substitute $\dfrac{\sin θ}{\cos θ}$ for tan θ in the given equation.

sin θ = –8/15 cos θ

Multiply both sides by cos θ.

Now that we have sin θ defined in terms of cos θ, let's use the Pythagorean trigonometric identity.

$\sin^2 θ + \cos^2 θ = 1$

$(–8/15 \cos θ)^2 + \cos^2 θ = 1$

Substitute –8/15 cos θ for sin θ.

$64/225 \cos^2 θ + \cos^2 θ = 1$

Square the entire expression (–8/15 cos θ).

$289/225 \cos^2 θ = 1$

Combine the like terms on the left side of the equation.

$\cos^2 θ = 225/289$

$\cos θ = ± \sqrt{\dfrac{225}{289}}$

Multiply both sides of the equation by 225/289.
Take the square root of both sides.

cos θ = ±15/17   Evaluate the square root.

The cosine of angles between 3π/2 and 2π is positive, so cos θ = 15/17.

20

Lesson 4.6
Trigonometric Functions in Real Life

There are many actions that complete some sort of regular cycle periodically and can be modeled by trigonometric functions. For example, the swinging arc of a pendulum provides a relationship of spatial location to time that can be described by a sine or cosine function.

EXAMPLE 21

Rebecca is preparing for her trapeze performance. She swings on a trapeze that is connected by 20-foot ropes to its pivot point. She begins her swing from a platform, where the angle of the ropes from the vertical is 2/5 π radians, as shown in the diagram below, and swings to the opposite side and back, repeatedly.

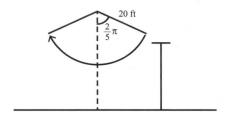

It takes her 8.8 seconds to make one full swing back and forth, returning to her original position.

Let us call the equilibrium position, where Rebecca, on her trapeze, is hanging directly below the pivot point, the zero position. (Note that she only passes through this point during her swings and does not hang stationary there.) This is represented by the point where the arc intersects the vertical dashed line in the diagram. We'll measure her displacement from this position along the arc she travels, with positive numbers representing locations to the right of the zero position and negative numbers representing locations to the left. The absolute value of her displacement at any time is the distance she is along the arc from the equilibrium position.

(a) Write and graph a trigonometric function that represents Rebecca's displacement, in feet, from the equilibrium position as a function of time, in seconds.

(b) There is a 20-second musical introduction, at the end of which Rebecca must be at the opposite extreme of her swing from her beginning position. If she wants to swing for as much of the musical introduction as possible, how long should she wait after the music begins to start swinging?

That's a wordy question! Let's begin by solving for Rebecca's displacement measurement at her starting position, which is also the extreme of her swing to the right of the equilibrium position. Remember, $\theta = s/r$, where θ is the central angle measure, s is the arc length subtended by that angle, and r is the radius. For Rebecca's starting position, when she's on the platform, $\theta = 2/5 \pi$ and $r = 20$.

$2/5 \pi = s/20$	Substitute $2/5 \pi$ for θ and 20 for r.
$8\pi = s$	Multiply both sides by 20.
$s \approx 25$	Substitute 3.14 as an approximation for π and multiply by 8.

Let d represent Rebecca's displacement from equilibrium, as described above. Her maximum displacement along the arc is 25 feet ($d = 25$). The extreme point on her swing to the left is also an arc length of 25 feet from the zero position, but we are defining points to the left of center as negative, so $d = -25$ at the left extreme. Rebecca's displacement, d, ranges from -25 to 25.

> We can assume that Rebecca's swing goes approximately the same distance in each direction, like a pendulum.

We must write a trigonometric function (sine or cosine) that relates Rebecca's displacement to time. She completes a full cycle (swinging from her beginning position to the opposite extreme and back again to her original position) in 8.8 seconds, so the period of this trigonometric function is 8.8. Her maximum displacement is 25 and her minimum displacement is -25. So, the midline is $d = 0$, and the amplitude is 25.

Let t represent time, in seconds, since Rebecca started her swing from the platform. At $t = 0$, $d = 25$, because she is at her right-side extreme displacement. At $t = 8.8$, $d = 25$ again, at the completion of her swing back and forth. Because this is a sine or cosine function, her swing across and her swing back take the same amount of time, so she reaches $d = -25$ halfway through the period, at $t = 4.4$. She also crosses the vertical equilibrium line, where $d = 0$, halfway through each swing, at $t = 2.2$ and $t = 6.6$.

Use this information to graph a curve representing Rebecca's ongoing motion swinging back and forth on her trapeze.

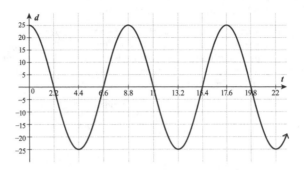

This function starts at its maximum when $t = 0$, so let's use a cosine function to describe the graph. This is a cosine function with a midline of 0, an amplitude of 25, and a period of 8.8. The frequency is different from that of a standard cosine function, which has a period of 2π radians. Remember, the period of a cosine function $y = a \cos b(x − c) + d$ is $2\pi/b$, so we can set this equal to 8.8 to solve for b.

$8.8 = 2\pi/b$

$8.8b = 2\pi$	Multiply both sides by b.
$b = 2\pi/8.8$	Divide both sides by 8.8.
$b = 20\pi/88$	Multiply both sides by 10 to have a whole number in the denominator.
$b = 5\pi/22$	Divide the numerator and denominator by 4 to simplify the fraction.

It is not necessary to convert the fraction to be a ratio of whole numbers, but it makes the equation cleaner and easier to work with.

So, the equation that represents Rebecca's displacement, in feet, from the equilibrium position, as a function of time, in seconds, is $d = 25 \cos 5\pi/22\ t$.

We could also use function $d = 25 \sin 5\pi/22\ (t + 2.2)$, a sine function of the same amplitude, midline, and period but shifted 2.2 units to the left.

It takes Rebecca 4.4 seconds to reach the opposite extreme of her swing after leaving the platform. She will return to that position again after a full cycle, 8.8 seconds later, which is a total of 13.2 seconds after leaving the platform (at $t = 13.2$). And, she will again return to that position another 8.8 seconds later, at $t = 22.0$.

The musical introduction is only 20 seconds long, so she can't take 22 seconds to get there. In other words, she doesn't have enough time for $2\frac{1}{2}$ cycles of swinging back and forth. She can either swing a 1/2 cycle (going directly from the platform across to the opposite extreme) or $1\frac{1}{2}$ cycles (going across, coming back, and going across again). She wants to swing for as much of the musical introduction as possible, so she should swing $1\frac{1}{2}$ cycles, which takes a total of 13.2 seconds. To arrive there right at the end of the 20-second musical introduction, she must wait for (20 – 13.2) seconds, or 6.8 seconds, after the music begins before starting to swing.

You can also find this information in the graph. The points at which the graph reaches its minimum, where $d = -25$, are when $t = 4.4$, $t = 13.2$, and $t = 22.0$.

CHAPTER 4 PRACTICE QUESTIONS

Directions: Complete the following open-ended problems as specified by each question stem. For extra practice after answering each question, try using an alternative method to solve the problem or check your work.

1. Using the triangle diagram below, prove that $\tan(90 - \theta) = \dfrac{1}{\tan \theta}$, where θ is the indicated angle measure, in degrees.

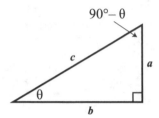

2. Given $\csc \theta = -4/3$ and $3\pi/2 < \theta < 2\pi$, determine the values of the other five trigonometric functions for angle θ. Rationalize the denominators, if needed.

3. The central angle for the subtended arc of the circle $x^2 + y^2 = 144$ is $225°$. Find the following:

   (a) The sine and cosine of the angle

   (b) The length of the arc, in radians

4. For each of the following, sketch the graph of the function. Use appropriate scales for axes:

   (a) $y = 2 \sin \left(x - \dfrac{\pi}{3}\right)$

   (b) $y = 4 - \tan \dfrac{2x}{3}$

   (c) $y = -2 \cos (3x + \pi)$

5. Prove each of the following identities, then demonstrate that the identity holds true for the triangle shown below.

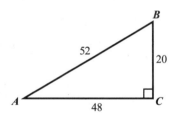

   (a) $\sin \theta \times \sec \theta = \tan \theta$

   (b) $\dfrac{\cos \theta}{\csc \theta} = \dfrac{\sin \theta}{\sec \theta}$

   (c) $1 + \cot^2 \theta = \csc^2 \theta$

6. Write an equation that describes the trigonometric function graphed below.

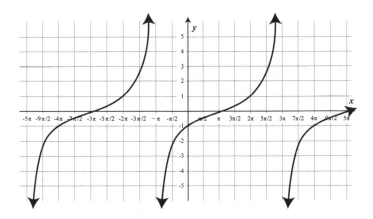

7. Shane plays a note of A on his French horn. The sound wave he creates has a frequency of 440 hertz (cycles per second) and an amplitude of 0.5. Assuming a midline of the x-axis, what equation describes the sound wave Shane created?

8. Amy runs around a circular track near a three-story building. The closest Amy comes to the building while running is 6 meters, and this is her starting point. She runs at a steady pace, completing one lap, which is 402 meters long, every 80 seconds, for a total of 5 laps.

(a) Write and graph a trigonometric function that represents Amy's direct distance from the building as a function of time, in seconds.

(b) The building's shadow extends 90 meters from the base of the building, shading all of the track that is within this distance of the building. The remainder of the track is in direct sunlight. For approximately how many seconds out of each lap is Amy running in direct sunlight?

SOLUTIONS TO CHAPTER 4 PRACTICE QUESTIONS

1. **See proof below.**
   If you need to remind yourself of the relationship of each side of the triangle to a given angle, you can sketch and label two triangles, as shown. Pay attention to the labeling of the different angles in question. The triangles are identical, but the angle position changes the relation of the sides when producing the trigonometric ratios.

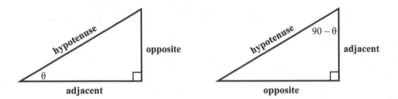

   Using the given side lengths a, b, and c, evaluate the required trig functions. You are trying to

   show that tan (90° – θ) is equal to 1/tan θ, so if you substitute side measures in for each, you'll

   get $\dfrac{b}{a} = \dfrac{1}{\dfrac{a}{b}}$. Multiply by the reciprocal on the right side and there's your identity.

2. **sin θ = –3/4, csc θ = –4/3, cos θ = $\dfrac{\sqrt{7}}{4}$, sec θ = $\dfrac{4\sqrt{7}}{7}$, tan θ = $\dfrac{-3\sqrt{7}}{7}$, cot θ = – $\dfrac{\sqrt{7}}{3}$**

   Recall that cosecant is the reciprocal of the sine function, so sin θ = –3/4. One down, four more to go. Sketch a right triangle in the coordinate plane. The measure of θ is between $3\pi/2$ and 2π, so the terminal side is in Quadrant IV. The side opposite θ is 3 units long (with a negative value, because it is below the x-axis), and the hypotenuse (the terminal side) is 4 units long. Using the Pythagorean theorem ($a^2 + b^2 = c^2$), you can find the length of the third side, which lies on the x-axis.

$(3)^2 + b^2 = (4)^2$	Substitute given values.
$9 + b^2 = 16$	Evaluate expression.
$b^2 = 7$	Subtract 9 from both sides.
$b = \sqrt{7}$	Take the square root of both sides.

The leg adjacent to the enclosed angle lies along the *x*-axis to the right of the origin, so it has a positive measure. Now we can sketch the right triangle.

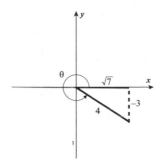

Evaluate the other four functions given these values.

Known: $\sin \theta = -\dfrac{3}{4}$ $\csc \theta = -\dfrac{4}{3}$

Remaining: $\cos \theta = \dfrac{\sqrt{7}}{4}$ $\sec \theta = \dfrac{4}{\sqrt{7}} \times \dfrac{\sqrt{7}}{\sqrt{7}} = \dfrac{4\sqrt{7}}{7}$

$\tan \theta = \dfrac{-3}{\sqrt{7}} \times \dfrac{\sqrt{7}}{\sqrt{7}} = \dfrac{-3\sqrt{7}}{7}$ $\cot \theta = -\dfrac{\sqrt{7}}{3}$

3. **(a) $(-\sqrt{2})/2$; (b) 15π**

    (a) Draw an image to get an idea of what is going on and create a right triangle with the *x*-axis:

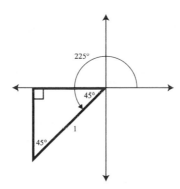

225° opens up to Quadrant III, and when the triangle is created with the *x*-axis, a 45° angle is created. This makes it an isosceles right triangle, or 45-45-90 special right triangle, with a hypotenuse of 1 in the unit circle. The side lengths of the triangle can be determined using the ratio of $s\colon s\colon s\sqrt{2}$. This ratio produces the lengths of the base and height of $1/\sqrt{2}$. Rationalize the denominator by multiplying both numerator and denominator by $\sqrt{2}$.

Note the location of where the triangle is drawn, Quadrant III. In Quadrant III, both the *x*- and *y*-coordinates are negative. Label the figure:

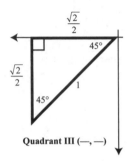

Quadrant III (—, —)

Now, determine the sine and cosine using SOHCAHTOA.

$\sin\theta = \dfrac{opposite}{hypotenuse}$ 　　　　　　　　　Definition of sine.

$\sin 225° = \dfrac{\dfrac{-\sqrt{2}}{2}}{1}$ 　　　　　　　　Substitute calculated values.

$\sin 225° = -\sqrt{2}/2$ 　　　　　　　　　Simplify.

$\cos\theta = \dfrac{adjacent}{hypotenuse}$ 　　　　　　　　　Definition of cosine.

$\cos 225° = \dfrac{\dfrac{-\sqrt{2}}{2}}{1}$ 　　　　　　　　Substitute calculated values.

$\cos 225° = -\sqrt{2}/2$ 　　　　　　　　　Simplify.

(b) Now, convert to radians by multiplying the given angle (225°) by the conversion proportion of π radians/180°. This yields $5/4\,\pi$ radians.

A circle described by the equation $x^2 + y^2 = 144$, or $x^2 + y^2 = 12^2$, has a radius of 12. The radian measure of a central angle is equal to s/r, where s is the arc length of the subtended arc and r is the radius. Set up an equation using the radius and the radian measure we found.

$5/4\,\pi = s/12$
$12(5/4\,\pi) = s$ Multiply both sides by 12.
$15\pi = s$

A $5/4\,\pi$ central angle subtends an arc of length 15π in the circle $x^2 + y^2 = 144$.

4. See the following graphs.

(a) Determine amplitude, period, and any necessary shifts:

Amplitude = $|a| = |2| = 2$

Period = $\dfrac{2\pi}{b} = \dfrac{2\pi}{1} = 2\pi$

Shift: Shifted to the right $\dfrac{\pi}{3}$ units

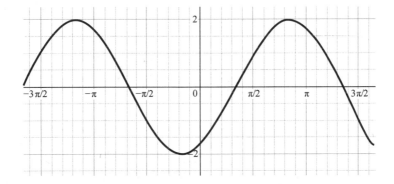

(b) Determine amplitude, period, and any necessary shifts:

Amplitude = $|a| = |-1| = 1$

The negative a value means that the graph is reflected across its midline.

Remember, the parent tangent function has a period of π, not 2π.

Period = $\dfrac{\pi}{b} = \dfrac{\pi}{\frac{2}{3}} = \pi \times \dfrac{3}{2} = 3\pi/2$

Shift: Shifted up 4 units

Remember to reflect the tangent function across the midline, $y = 4$.

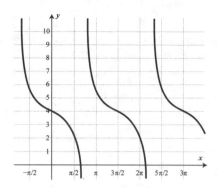

(c) Determine amplitude, period, and any necessary shifts:

Amplitude = $|a| = |-2| = 2$

The negative a value means that the graph is reflected across its midline, which in this case is the x-axis.

Period = $\dfrac{2\pi}{b} = \dfrac{2\pi}{3}$

Shift: Shifted to the left π units

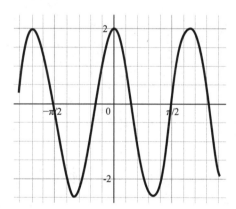

5. **See the following proofs.**

(a) $\sin \theta \times \sec \theta = \tan \theta$ Given.

$\sin \theta \times \dfrac{1}{\cos \theta} = \tan \theta$ Identity $\sec \theta = \dfrac{1}{\cos \theta}$.

$\dfrac{\sin \theta}{\cos \theta} = \tan \theta$ Multiply terms on left side of equation.

$\tan \theta = \tan \theta$ Identity $\tan \theta = \dfrac{\sin \theta}{\cos \theta}$.

Now, substitute appropriate side lengths from the given triangle to demonstrate the identity.

$\sin\theta \times \sec\theta = \tan\theta$	Given.
$(\dfrac{20}{52})(\dfrac{52}{48}) = \dfrac{20}{48}$	SOHCAHTOA and reciprocal identities.
$\dfrac{20}{48} = \dfrac{20}{48}$	Simplification.

(b)

$\dfrac{\cos\theta}{\csc\theta} = \dfrac{\sin\theta}{\sec\theta}$	Given.
$\cos\theta \div \csc\theta = \sin\theta \div \sec\theta$	Write fractions as divisions.
$\cos\theta \div \dfrac{1}{\sin\theta} = \sin\theta \div \dfrac{1}{\cos\theta}$	Identities $\csc\theta = \dfrac{1}{\sin\theta}$ and $\sec\theta = \dfrac{1}{\cos\theta}$.
$\cos\theta \times \sin\theta = \sin\theta \times \cos\theta$	Division is equal to multiplication by reciprocal.
$\cos\theta \times \sin\theta = \cos\theta \times \sin\theta$	Commutative property of multiplication.

Now, substitute appropriate side lengths from the given triangle to demonstrate the identity.

$\dfrac{\cos\theta}{\csc\theta} = \dfrac{\sin\theta}{\sec\theta}$	Given.
$\dfrac{\frac{48}{52}}{\frac{52}{20}} = \dfrac{\frac{20}{52}}{\frac{52}{48}}$	SOHCAHTOA and reciprocal identities.
$(\dfrac{48}{52})(\dfrac{20}{52}) = (\dfrac{20}{52})(\dfrac{48}{52})$	Rewrite division as multiplication by reciprocal.

(c) $1 + \cot^2 \theta = \csc^2 \theta$ Given.

$$1 + \left(\frac{\cos \theta}{\sin \theta}\right)^2 = \left(\frac{1}{\sin \theta}\right)^2$$ Identities $\cot \theta = \dfrac{\cos \theta}{\sin \theta}$ and $\csc \theta = \dfrac{1}{\sin \theta}$.

$$1 + \frac{\cos^2 \theta}{\sin^2 \theta} = \frac{1}{\sin^2 \theta}$$ Square terms.

$\sin^2 \theta + \cos^2 \theta = 1$ Multiply both sides of equation by $\sin^2 \theta$.

$1 = 1$ Identity $\sin^2 \theta + \cos^2 \theta = 1$.

Now, substitute appropriate side lengths from the given triangle to demonstrate the identity.

$1 + \cot^2\theta = \csc^2\theta$ Given.

$$1 + (\frac{48}{20})^2 = (\frac{52}{20})^2$$ Reciprocal identities.

$$1 + \frac{2304}{400} = \frac{2704}{400}$$ Perform arithmetic.

$$\frac{400}{400} + \frac{2304}{400} = \frac{2704}{400}$$ Put first fraction in common denominator form.

$$\frac{2704}{400} = \frac{2704}{400}$$ Add numerators to produce identity statement.

6. **$y = \tan 1/4\,(x - \pi)$**
    This function has asymptotes and is not a continuous wave, so it is a tangent function. The parent tangent function has a period of π, but this function has a much longer period. It has x-intercepts of -3π, π, and 5π, which are each 4π units apart. It completes a full cycle in 4π units, so it has a period of 4π. Find the value of b in $y = a \tan b(x - c) + d$ using the fact that the period of a tangent function equals π/b.

    $4\pi = \pi/b$

    $4\pi b = \pi$ Multiply both sides by b.

    $b = 1/4$ Divide both sides by 4π and cancel the common factor of π.

    The points of inflection lie on the x-axis, so there is no vertical shift, and $d = 0$ in $y = a \tan b(x - c) + d$. There is, however, a horizontal shift. The function $y = \tan x$ passes through the origin, but the function shown in this graph crosses at $x = \pi$ instead. It is shifted π units to the right, so $c = \pi$. So far, we can write the equation as $y = a \tan 1/4 \, (x - \pi)$. To find the value of a, use the coordinates of a point on the graph in this equation. Let's use the point $(2\pi, 1)$.

$1 = a \tan 1/4 \, (2\pi - \pi)$	Substitute 2π for x and 1 for y.
$1 = a \tan \pi/4$	Simplify $1/4 \, (2\pi - \pi)$.
$1 = a \cdot 1$	Evaluate $\tan \pi/4$.
$1 = a$	Multiply.

The value of a is 1, so an equation that describes the graphed function is $y = \tan 1/4 \, (x - \pi)$.

7. **$y = 0.5 \sin 880\pi x$**
A sound wave could be described by a sine or a cosine function, but sine is more commonly used. The frequency of this sound wave is 440 cycles per second, so a single wave is completed in 1/440 second. The period of the sine wave is 1/440 second. The period of a sine function of the form $y = a \sin b(x - c) + d$ is equal to $2\pi/b$. Set $2\pi/b$ equal to 1/440 to solve for b.

$2\pi/b = 1/440$

$b = 880\pi$ \qquad\qquad Cross-multiply.

With a midline of the x-axis, there is no vertical shift. Where we begin the sine curve on the x-axis does not matter in this situation, so we can begin at the origin, with no horizontal shift of the sine curve. Without any shifts, the function is of the form $y = a \sin bx$. The amplitude is equal to a, so $a = 0.5$. Substitute this and the value we found for b. Shane's note of A produced a sound wave described by $y = 0.5 \sin 880\pi x$.

8. **(a) $f(x) = -64 \cos \pi/40 \, x + 70$, (b) 32 seconds**
(a) It may help to draw a sketch to represent the situation. This is only a rough sketch, because we do not know how much of the track the shadow covers, and nothing is to scale.

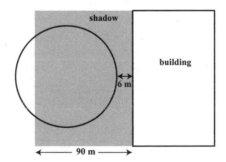

Because Amy is traveling along a circle, her direct distance to the side of the building (a straight line) can be described by a sine or cosine function in terms of the central angle associated with his position. Because she is moving at a steady pace, the central angle varies directly with time in seconds. So, we can write a function for $f(x)$, her direct distance, in meters, from the building, where x is time, in seconds.

At $x = 0$, when Amy starts her run, she is 6 meters from the building, so $f(0) = 6$. This is her minimum distance from the building during her laps, so we'll use a cosine curve reflected across its midline.

Amy completes one lap of the track in 80 seconds, so the cosine curve has a period of 80 seconds. The period of a cosine function is $2\pi/b$, where b is the coefficient of x in the function, so $2\pi/b = 80$. Multiplying both sides by b gives us $80b = 2\pi$, so $b = \pi/40$.

The circumference of Amy's circular path on the track is 402 meters, and the circumference of a circle is $2\pi r$, so $2\pi r = 402$, and $r = 201/\pi$, which is approximately equal to 64. The radius of the circular track is 64 meters. So, the diameter of the track is 128 meters. Amy's furthest point from the building is a distance of the diameter plus 6 meters: $128 + 6 = 134$. The cosine curve will have a minimum $f(x)$-value of 6 and a maximum $f(x)$-value of 134, so its midline will be at their average: $(6 + 134) \div 2 = 70$. Its amplitude is the difference between the midline and each extreme ($134 - 70$ or $70 - 6$), 64.

We'll write the function in the form $f(x) = a \cos b(x - c) + d$. The amplitude, $|a|$, is 64, but to reflect the cosine curve across the midline, a must be negative: -64. The b-value is $\pi/40$. There is no horizontal shift, so $c = 0$. The midline is 70, so the vertical shift, or d, is 70. The function that describes Amy's direct distance, in meters, from the building as a function of time, in seconds, is $f(x) = -64 \cos \pi/40 \, x + 70$. Its graph is below, showing 5 full cycles for Amy's 5 laps around the track.

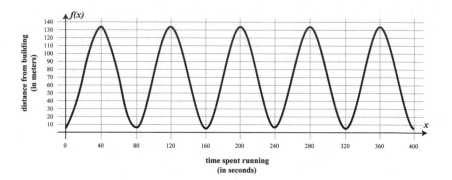

(b) The shadow extends 90 meters from the base of the building, or to $f(x) = 90$. The portions of the cosine curve where $f(x) > 90$ correspond to when Amy is in direct sunlight. We can find the x-values in the first cycle (between 0 and 80) where $f(x) = 90$, then subtract to find the time difference.

$-64 \cos \pi/40 \, x + 70 = 90$	Set cosine function equal to 90.
$-64 \cos \pi/40 \, x = 20$	Subtract 70 from both sides.
$\cos \pi/40 \, x = -0.3125$	Divide both sides by -64.
$\pi/40 \, x \approx 1.89$	Use your calculator to find $\cos^{-1}$ in radians.
$x \approx 24$	Multiply both sides by $40/\pi$.

We know the cosine curve is symmetrical with respect to any vertical line through a maximum or minimum, so it is symmetrical with respect to $x = 40$. The point where the curve first reaches 90 is when $x = 24$, which is 16 units to the left of 40, so it will come back down to 90 the same distance to the right of 40: $40 + 16 = 56$. Amy will be in direct sunlight from $x = 24$ to $x = 56$, which is a time period of $56 - 24 = 32$. Amy runs in direct sunlight for 32 seconds out of each lap.

REFLECT

Congratulations on completing Chapter 4!
Here's what we just covered.
Rate your confidence in your ability to

- Use the relationship between radius, radian measure, and arc length to solve for any one of these values for a given circle and central angle

  ① ② ③ ④ ⑤

- Convert between degrees and radians for any angle measure

  ① ② ③ ④ ⑤

- Find the value of the sine, cosine, tangent, cosecant, secant, or cotangent, where it exists, for any real number

  ① ② ③ ④ ⑤

- Graph sine, cosine, and tangent functions using information about the period, midline, amplitude, horizontal shift, and any reflection, and write an equation to describe a given sine, cosine, or tangent function

  ① ② ③ ④ ⑤

- Prove and apply trigonometric function identities

  ① ② ③ ④ ⑤

- Write and use sine and cosine functions to represent real-life situations and solve problems

  ① ② ③ ④ ⑤

If you rated any of these topics lower than you'd like, consider reviewing the corresponding lesson before moving on, especially if you found yourself unable to correctly answer one of the related end-of-chapter questions.

 Access your online student tools for a handy, printable list of Key Points for this chapter. These can be helpful for retaining what you've learned as you continue to explore these topics.

Chapter 5
Logarithms

GOALS By the end of this chapter,
you will be able to

- Find the inverse relation of a given function, including exponential functions, and determine any necessary domain restrictions on the function to make its inverse a function

- Evaluate or estimate numerical logarithms, where they exist

- Solve single-variable equations containing logarithmic expressions

- Use binary, natural, and common logarithms

- Use logarithmic identities

- Graph logarithmic functions

- Write and use logarithmic functions to represent real-life situations and solve problems

5

Lesson 5.1
Inverse Functions

REVIEW

A **relation** is set of ordered pairs, usually of the form (x, y), or an equation that describes a relationship between values in two sets. Functions are relations, but not all relations are functions.

Vertical line test: For a graph of a relation, if you can draw a vertical line somewhere that intersects the graph more than once, then the relation is **not** a function.

Switching x and y in a given relation reflects the graph of the relation across the line $y = x$.

For a refresher on domains and ranges you'll want to refer back to Lesson 3.3. Inverse operations can be found in Lesson 3.4.

A function maps each element in its domain (each x-value) to a single corresponding element in its range (a y-value, or $f(x)$-value). An **inverse function** takes every element in the original function's range and maps it back to its corresponding element in the original function's domain. So, the domain of a function is the range of its inverse function, and the range of a function is the domain of its inverse function.

In other words, an inverse function "undoes" or "reverses" whatever the original function did. In the simplest cases, an inverse function uses the inverse operation of the one used in the original function.

The functions $f(x) = x - 8$ and $g(x) = x + 8$ are inverse functions. The functions $p(x) = x/3$ and $q(x) = 3x$ are also inverse functions. If you take an x-value in the domain of $f(x)$, such as 10, the function $f(x)$ produces the value 2. The inverse function, $g(x)$, takes 2 and produces the value 10. The inverse function undid the work of $f(x)$ and restored the original number. Likewise, $p(x)$ takes an x-value of 12 and produces a $p(x)$-value of 4, while $q(x)$ takes an x-value of 4 and produces a $q(x)$-value of 12.

When a function involves more than one operation, writing the inverse becomes a little more complicated. The idea is to create a new function that solves for each x-value from the original function's domain, given its $f(x)$-value, or solving for x in terms of y. The only catch is that the old y-values must be the new domain values in the inverse function, meaning they must be x-values.

The notation $f^{-1}(x)$ indicates the inverse function of $f(x)$. The −1 is not an exponent, so $f^{-1}(x)$ does **not** mean $[f(x)]^{-1}$.

To find an inverse for a function, switch the x and y variables and solve for the new y in terms of the new x.

This way, a coordinate pair (a, b) in the original function is reproduced as (b, a) in the inverse function.

EXAMPLE 1

If $f(x) = \dfrac{5}{x-2}$, what is $f^{-1}(x)$, its inverse?

Use y in place of $f(x)$, then switch the variables and solve for the new y.

$y = \dfrac{5}{x-2}$	Rewrite with y instead of $f(x)$.
$x = \dfrac{5}{y-2}$	Replace y with x and x with y.
$x(y-2) = 5$	Multiply both sides by $(y-2)$.
$xy - 2x = 5$	Distribute the multiplication through the binomial factor.
$xy = 2x + 5$	Add $2x$ to both sides.
$y = \dfrac{2x+5}{x}$	Divide both sides by x.
$y = 2 + 5/x$	Rewrite the fraction as two terms and simplify.

The inverse function is $f^{-1}(x) = 5/x + 2$.

Let's try finding the composite function $f^{-1}(f(x))$. A composite function is a function within a function. The outer function takes the output (function) values of the inner function and uses them as input values for its own function equation.

A composition of functions $f(x)$ and $g(x)$, specifically f composed with g of x, can be written as $f(g(x))$ or as $(f \circ g)(x)$. Likewise, g composed with f of x can be written as $g(f(x))$ or as $(g \circ f)(x)$.

In Example 1, $f(x) = \dfrac{5}{x-2}$ and $f^{-1}(x) = 5/x + 2$. To write the composite function $f^{-1}(f(x))$, replace $f(x)$ with $\dfrac{5}{x-2}$ and then use that as x in the $f^{-1}(x)$ equation.

$f^{-1}(f(x)) = f^{-1}\left(\dfrac{5}{x-2}\right)$ Substitute $\dfrac{5}{x-2}$ for $f(x)$.

$= \dfrac{5}{\dfrac{5}{x-2}} + 2$ Substitute $\dfrac{5}{x-2}$ for x in the $f^{-1}(x)$ equation.

$= 5 \cdot \dfrac{x-2}{5} + 2$ Rewrite division by $\dfrac{5}{x-2}$ as multiplication by the reciprocal, $\dfrac{x-2}{5}$.

$= x - 2 + 2$ Multiply and cancel out the common factor of 5.

$= x$ Add -2 and 2.

So, $f^{-1}(f(x)) = x$. This demonstrates that the inverse undoes the work of $f(x)$, leaving us with each original element of the domain of $f(x)$: each x-value.

Here is how you may see inverse functions on the ACT.

Given $p(x) = \sqrt[3]{x+7}$, which of the following expressions is equal to $p^{-1}(x)$ for all real numbers x ?

     A. $\sqrt[3]{x} - 7$
     B. $-\sqrt[3]{x+7}$
     C. $\sqrt[3]{x-7}$
     D. $(x+7)^3$
     E. $x^3 - 7$

What about function f composed with its inverse function, or $f(f^{-1}(x))$?

$f(f^{-1}(x)) = f(5/x + 2)$ Substitute $5/x + 2$ for $f^{-1}(x)$.

$$= \frac{5}{\left(\dfrac{5}{x} + 2\right) - 2}$$ Substitute $5/x + 2$ for x in the $f(x)$ equation.

$$= \frac{5}{\dfrac{5}{x}}$$ Combine 2 and -2 in the denominator.

$$= 5 \cdot \frac{x}{5}$$ Rewrite division by $5/x$ as multiplication by the reciprocal, $x/5$.

$$= x$$ Multiply and simplify.

So, $f(f^{-1}(x))$ is also equal to x. A function and its inverse undo one another. It works in either direction.

> Functions $f(x)$ and $g(x)$ are inverses of one another if and only if $f(g(x)) = x$ and $g(f(x)) = x$ for all values of x in their domains.

Some functions have an inverse relation that is not a function. Sometimes a function maps more than one x-value to the same y-value. Its inverse relation must then map that one function value back to two different values, which means that it breaks the rules of function behavior.

EXAMPLE 2

Given $y = x^2 - 3$, find its inverse relation, and graph both on the same coordinate grid.

Switch the variables x and y, and solve for the new y.

$$x = y^2 - 3$$
$$y^2 = x + 3$$
$$y = \pm \sqrt{x + 3}$$

The inverse relation is $y = \pm \sqrt{x + 3}$. This is not a function, because it produces both a positive and a negative number for each x-value (other than -3, which produces just 0). The graphs of $y = x^2 - 3$ and $y = \pm \sqrt{x + 3}$ are shown on the coordinate grid below.

Notice that the graph of $y = \pm \sqrt{x + 3}$ fails the vertical line test for a function. A vertical line can be drawn that passes through this graph in more than one place.

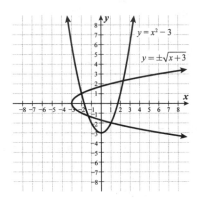

$$y = \sqrt[3]{x + 7}$$ Write using y instead of $p(x)$.

$$x = \sqrt[3]{y + 7}$$ Switch x and y.

$$x^3 = y + 7$$ Cube both sides.

$$x^3 - 7 = y$$ Subtract 7 from both sides to isolate y.

So, $p^{-1}(x) = x^3 - 7$. The correct answer is (E).

Notice that, together, the two graphs are symmetrical with respect to the diagonal line $y = x$. This makes sense, because the way to reflect a graph across the line $y = x$ is by switching the variables x and y in the given equation. We switched x and y in the equation $y = x^2 - 3$, producing a relation ($y = \pm\sqrt{x+3}$) that is the reflection of the original function across the line $y = x$.

Although $y = \pm\sqrt{x+3}$ is not a function, if we include only the top half of its parabola, represented by the equation $y = \sqrt{x+3}$, then it becomes a function. In this case, though, only half of the function $y = x^2 - 3$ is its inverse. We are restricting the range of the relation $y = \pm\sqrt{x+3}$ to only positive y-values and 0 (or $y \geq 0$). The range of the inverse is the domain of the original function, so the domain of $y = x^2 - 3$ must be restricted to only $x \geq 0$. The inverse functions $y = x^2 - 3$ for $x \geq 0$ and $y = \sqrt{x+3}$ are shown on the graph below.

Alternatively, we could have restricted the range of $y = \pm\sqrt{x+3}$ to only negative numbers and 0. The functions $y = x^2 - 3$ for $x \leq 0$ and $y = -\sqrt{x+3}$ are inverse functions.

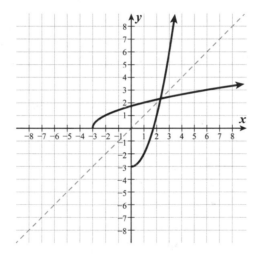

This graph also shows the line of reflection, $y = x$, as a dashed line.

Lesson 5.2
Logarithms

A logarithm is the inverse operation to exponentiation (raising a number to an exponent). The logarithm undoes the work of the exponential expression.

The term "log" is short for "logarithm," and the b in $\log_b y$ is its base, just as b is the base in b^x.

> If $y = b^x$, then $\log_b y = x$. Likewise, if $\log_b y = x$, then $y = b^x$.

So, the logarithm is the exponent to which the base must be raised to produce the value y in $\log_b y$. For example, $\log_4 64$ is the exponent to which 4 must be raised to produce 64. The equation $\log_4 64 = x$ can be rewritten as $4^x = 64$. Because 4 cubed is equal to 64, $4^x = 4^3$, and $x = 3$.

If the same base is raised to an exponent on either side of the equation (without any other terms or factors), then the exponents are equal: If $b^x = b^y$, then $x = y$.

Evaluate each of the following expressions, where they exist.

$\log_2 8$	$\log_4 1/4$
$\log_9 81$	$\log_5 1/25$
$\log_3 81$	$\log_8 1$
$\log_{10} 100{,}000$	$\log_2 -4$
$\log_7 7$	$\log_6 0$

Set each logarithm equal to x and rewrite the equation in exponential form, then solve.

$\log_2 8 = x$

$2^x = 8$
$2^x = 2^3$
$x = 3$, so $\log_2 8 = 3$

$\log_9 81 = x$

$9^x = 81$
$9^x = 9^2$
$x = 2$, so $\log_9 81 = 2$

$\log_3 81 = x$

$3^x = 81$
$3^x = 3^4$
$x = 4$, so $\log_3 81 = 4$

$\log_{10} 100{,}000 = x$

$10^x = 100{,}000$
$10^x = 10^5$
$x = 5$, so $\log_{10} 100{,}000 = 5$

$\log_7 7 = x$

$7^x = 7$
$7^x = 7^1$
$x = 1$, so $\log_7 7 = 1$

$\log_4 1/4 = x$

$4^x = 1/4$
$4^x = 4^{-1}$
$x = -1$, so $\log_4 1/4 = -1$

$\log_5 1/25 = x$

$5^x = 1/25$
$5^x = 5^{-2}$
$x = -2$, so $\log_5 1/25 = -2$

$\log_8 1 = x$

$8^x - 1$
$8^x = 8^0$
$x = 0$, so $\log_8 1 = 0$

> The number 1 could be rewritten as any base to the power of 0, such as 5^0, but for our solution process 8^0 is most helpful.

$\log_2 -4 = x$

$2^x = -4$

There is no power of 2 that would produce -4. The expression $\log_2 -4$ is undefined.

$\log_6 0 - x$

$6^x = 0$

> If you are tempted to try letting $x = 0$, remember that $6^0 = 1$, not 0.

There is no power of 6 that results in 0. The expression $\log_6 0$ is undefined.

For any positive real number b, where $b \neq 1$:

$\log_b b = 1$
$\log_b 1 = 0$
$\log_b 0$ is undefined.
$\log_b c$ is undefined if c is negative.
$\log_b b^n = n$
$\log_b (1/b^n) = -n$

In the last step of solving each of the equations we wrote for Example 3, we set the exponents of matching bases equal to one another: If $b^x = b^y$, then $x = y$. However, what we were really doing at each of those moments was taking the logarithm of both sides of the equation.

Remember, the logarithm is the inverse operation to exponentiation. When solving an equation, you typically apply the inverse operation of those already applied to the variable. When you see $x + 4 = 9$, you use subtraction (of 4) on both sides, because it is the inverse of the addition shown. Likewise, when you see $3^x = 3^7$, you should take the logarithm (base 3) of both sides, to undo the exponentiation.

$b^x = b^y$

$\log_b b^x = \log_b b^y$ Take the logarithm with base b of both sides.

$x = y$ Use the rule $\log_b b^n = n$.

It's important to use the same base on both sides when taking the logarithm of both sides of an equation. As with any other operation applied to an equation, you must do exactly the same thing to both sides, to maintain the equality.

In Example 3, all logarithms that existed had integer values. However, exponents are sometimes fractions.

EXAMPLE

Find the value of each of the following expressions.

$$\log_{16} 4 \qquad\qquad \log_9 27$$

If $\log_{16} 4 = x$, then $16^x = 4$. The square root of 16 is 4: $\sqrt{16} = 4$, which can also be written as $16^{1/2} = 4$. Substitute $16^{1/2}$ for 4 in $16^x = 4$.

$16^x = 16^{1/2}$
$x = 1/2$ Take the logarithm, base 16, of both sides.

So, $\log_{16} 4 = 1/2$.

If $\log_9 27 = x$, then $9^x = 27$. We can write both sides of this equation as exponential expressions with base 3.

$(3^2)^x = 3^3$	Rewrite using a base of 3 on both sides.
$3^{2x} = 3^3$	Raise 3^2 to the power of x by multiplying the exponents.
$2x = 3$	Take the logarithm, base 3, of both sides.
$x = 3/2$	Divide both sides by 2 to solve for x.

So, $\log_9 27 = 3/2$.

4

To check our work, we can rewrite the equation $\log_9 27 = 3/2$ as $9^{3/2} = 27$. The numerator in the fractional exponent acts like a whole-number exponent, cubing 9. The denominator of 2 represents a square root. The expression $9^{3/2}$ can be written either as $\sqrt{9^3}$ or as $(\sqrt{9})^3$.

$\sqrt{9^3} = \sqrt{729} = 27$

$(\sqrt{9})^3 = 3^3 = 27$

As you become more familiar with logarithms, you may not need to write an equation with a variable (such as x) to evaluate numerical logarithms. Just ask yourself, "What exponent, when applied to that base, will give the number shown?" However, sometimes you must solve equations containing logarithms that already include one or more variables.

> The reason you can write the expression $9^{3/2}$ as either $(9^3)^{1/2}$ or $(9^{1/2})^3$ is because both of these represent 9 raised to the power of $(3 \cdot 1/2)$. By the commutative property of multiplication, $3 \cdot 1/2$ is equal to $1/2 \cdot 3$, so $9^{3 \cdot 1/2} = 9^{1/2 \cdot 3}$.

Here is how you may see logarithms on the ACT.

Which of the following ranges of consecutive integers contains the value of the expression $\log_8(8^{11/3})$?

A. 0 and 1
B. 1 and 2
C. 2 and 3
D. 3 and 4
E. 7 and 8

EXAMPLE 5

If $\log_{d+6} 625 = 4$, what is the value of d?

Write an equation using the definition of logarithm, then solve for d.

$(d + 6)^4 = 625$

$d + 6 = \sqrt[4]{625}$ You could instead write $(d + 6)^4 = 5^4$.

$d + 6 = 5$ Solve for $d + 6$.

$d = -1$ Subtract 6 from both sides.

EXAMPLE 6

Solve the following equation for x.

$$\log_2(x^2 - 2x) - 1 = 2$$

The parentheses tell us that the entire expression $x^2 - 2x$ is included in the logarithmic expression, whereas the −1 is not.

First, add 1 to both sides to put the equation in the form $\log_b m = n$, which can be rewritten as $b^n = m$. This notation is different from how we referred to this earlier ($\log_b y = x$ and $b^x = y$), but that's just because x is already being used. Any letter can be used as a variable, as long as it and its relationship to everything else is consistent throughout that equation.

$\log_2(x^2 - 2x) = 3$

$2^3 = x^2 - 2x$ Rewrite the relationship as an exponential equation.

$8 = x^2 - 2x$ Evaluate 2^3.

$0 = x^2 - 2x - 8$ Subtract 8 from both sides.

$0 = (x + 2)(x - 4)$ Factor the quadratic.

$x = -2$ or $x = 4$ Solve for each factor set equal to 0.

The solutions to $\log_2(x^2 - 2x) - 1 = 2$ are $x = -2$ and $x = 4$. If you substitute either value into the equation, you get $\log_2 8 - 1 = 2$, which simplifies to $3 - 1 = 2$, a true equation.

EXAMPLE 7

What is the value of _k_ in the equation below?

$$3^{\log_3 2k} = 9$$

We are given an exponential equation (with a funny looking exponent, but it is still just an exponent of 3), so we can use the definition of logarithm to rewrite the relationship as a logarithmic relationship.

$\log_3 9 = \log_3 2k$

$2 = \log_3 2k$	Simplify $\log_3 9$.	This is the same as taking the logarithm, base 3, of both sides of the equation $3^{\log_3 2k} = 9$.
$3^2 = 2k$	Rewrite as an exponential relationship, using the definition of logarithm.	
$9 = 2k$	Evaluate 3^2.	
$4.5 = k$	Divide both sides by 2.	

For any positive real number b, where $b \neq 1$, $b^{\log_b x} = x$.

Because the expression $\log_b x$ represents the exponent to which b must be raised to get x, raising b to that exponent is equal to x.

Using this identity as a shortcut in Example 7 would have brought us directly to the second-to-last line in the solution process, $2k = 9$.

The expression $\log_8(8^{11/3})$ is in the form $\log_b b^n$, so it is equal to n, the exponent 11/3.

$\log_8(8^{11/3}) = 11/3$

$\qquad = 3\dfrac{2}{3}$ \qquad Rewrite as a mixed number.

The value of $3\dfrac{2}{3}$ is between 3 and 4, so the correct answer is (D).

SPECIAL LOGARITHMS

Logarithms can have any positive number for a base, but those with bases of 2, e, or 10 are especially useful.

This number 2.718... probably seems rather random or arbitrary, but there are several ways in which e is mathematically meaningful. For example, it is the limit of $(1 + 1/n)^n$ as n approaches infinity.

A logarithm of base 2 is called a **binary logarithm** and is commonly used in computer science. Finding a binary logarithm means determining what power of 2 produces the given number. For example, $\log_2 32 = 5$ and $\log_2(1/8) = -3$.

A logarithm of base e is called a **natural logarithm**. The constant e is an irrational number with a value of about 2.718. Natural logarithms are usually written using the notation ln instead of log-base-e: $\log_e x = \ln x$.

EXAMPLE 🔒 8

Simplify the expression $\ln e^{2x}$.

The base is e, so by the definition of logarithm, $\ln e^{2x}$ is the exponent to which e must be raised in order to get e^{2x}. The answer is $2x$.

You can also view this as a version of the rule $\log_b b^n = n$: $\log_e e^{2x} = 2x$. Or, you can always go back to basics and use the definition of a logarithm to rewrite the expression in an exponential equation you can then solve.

$\ln e^{2x} = m$	Call the value of the expression m. We will solve for m.
$\log_e e^{2x} = m$	Rewrite the logarithm to remind yourself that it is of base e.
$e^m = e^{2x}$	Use the definition of logarithm to rewrite as an exponential relationship.
$m = 2x$	Set the exponents equal to one another. (Take the ln of both sides.)
$\ln e^{2x} = 2x$	Substitute $\ln e^{2x}$ for m.

A logarithm of base 10 is called a **common logarithm**. It is so common that if you see a logarithm written without a base, such as log5, you can generally assume that it is a common logarithm ($\log_{10} 5$ in this case). Common logarithms are often used to create logarithmic scales, such as the Richter scale.

A **logarithmic scale** uses logarithms of numbers instead of the numbers themselves for the breakdown of measurements, and it is especially helpful when representing a wide range of numbers. A difference of 1 unit on a logarithmic scale means multiplication or division by a factor of 10, so a 6.0-magnitude earthquake on the Richter scale is 10 times as strong as a 5.0-magnitude earthquake and 100 times as strong as a 4.0-magnitude earthquake.

EXAMPLE

A video game company posted the following profits for each of its first six years in business.

Year	Profits (in dollars)
1	2000
2	10,000
3	50,000
4	200,000
5	1,200,000
6	10,000,000

Graph the company's first six years' profits using a linear scale and using a logarithmic scale.

The profits range from 2000 to 10,000,000 dollars, so our y-axis should show values from 0 to at least 10,000,000. For a linear scale, the increments are evenly spaced and valued, so let's use tick marks of 1,000,000.

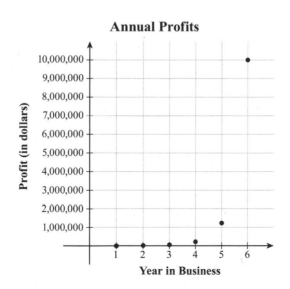

Although this gives some sense of the large growth in profits, it does not show the first few years' profits in much detail. If we plotted just the first three years with a linear scale of 0 to 50,000, we would have the graph below.

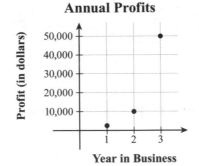

This graph clearly shows the relationship between the first three data points, but to include the other three data points using this scale, the graph would have to be 200 times as tall! We don't have enough paper for that. It's time to turn to a logarithmic scale.

A common logarithm is the exponent to which 10 must be raised to produce the given number. So, $\log_{10} 1000 = 3$ and $\log_{10} 10,000,000 = 7$. All of our data points will be between 3 and 7 using the common logarithm. Use your calculator to find each logarithmic value, rounded to the nearest tenth. This gives us the data points for this graph: (1, 3.3), (2, 4.0), (3, 4.7), (4, 5.3), (5, 6.1), and (6, 7.0). Here is a graph of the company's profits using a logarithmic scale.

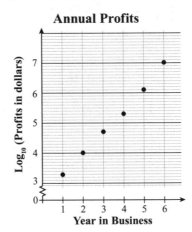

Graphs are often presented in a logarithmic scale but using the actual values rather than the log values as axis labels. For evenly spaced log intervals, the actual value intervals take up decreasing amounts of space between whole-number log values, as shown in the scale below.

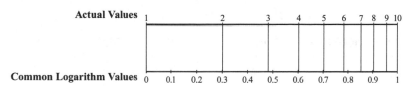

The logarithmic values are the exponents of 10 that produce the actual values. Evenly spaced increments of logarithmic values shown in our last graph of Annual Profits must be shown as intervals of varying sizes, as in the scale above, in order for us to label our y-axis with regular increments of actual profit values.

The x-axis of our graph is linear because the data points have a very simple set of evenly spaced x-values, {1, 2, 3, 4, 5, 6}. Sometimes graphs use a logarithmic scale on the horizontal x-axis, with either a linear or a logarithmic scale on the vertical axis. The best scales to use for representation depend on the given data set.

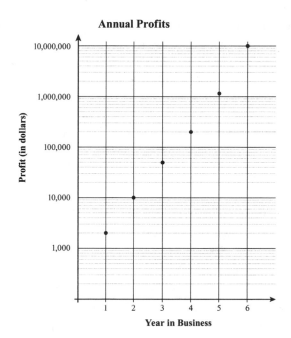

Lesson 5.3
Logarithmic Identities

Logarithmic identities are true equations that illustrate patterns in relationships between logarithms or parts of logarithms. We have already explored a number of logarithmic identities in this chapter: $\log_b b = 1$, $\log_b 1 = 0$, $\log_b b^x = x$, $\log_b(1/b^x) = -x$, and $b^{\log_b x} = x$.

There are also logarithmic properties related to mathematical operations. When you take a logarithm of a product, a quotient, a power, or a root, an interesting relationship emerges between the components and/or logarithms of the components.

Logarithmic Identities with Operations

For any positive real number b, where $b \neq 1$, any positive real numbers x and y, and any real number n:

Product property: $\log_b xy = \log_b x + \log_b y$

Quotient property: $\log_b(x/y) = \log_b x - \log_b y$

Power property: $\log_b x^n = n\log_b x$

Root property: $\log_b \sqrt[n]{x} = \dfrac{\log_b x}{n}$

EXAMPLE

Using $x = 8$, $y = 4$, $b = 2$, and $n = 3$, demonstrate that each of the four logarithmic identities shown above holds true.

Product property: $\log_b xy = \log_b x + \log_b y$

$\log_2(8 \cdot 4) = \log_2 8 + \log_2 4$	Substitute 8 for x, 4 for y, and 2 for b.
$\log_2 32 = \log_2 8 + \log_2 4$	Multiply $8 \cdot 4$.
$5 = 3 + 2$	Evaluate each of the three logarithms.
$5 = 5$	

Quotient property: $\log_b(x/y) = \log_b x - \log_b y$

$\log_2(8/4) = \log_2 8 - \log_2 4$	Substitute 8 for x, 4 for y, and 2 for b.
$\log_2 2 = \log_2 8 - \log_2 4$	Simplify 8/4.
$1 = 3 - 2$	Evaluate each of the three logarithms.
$1 = 1$	

Power property: $\log_b x^n = n\log_b x$

$\log_2 8^3 = 3\log_2 8$	Substitute 8 for x, 2 for b, and 3 for n.	For simpler arithmetic, we could have instead showed that $\log_b y^n = n\log_b y$: $\log_2 4^3 = 3\log_2 4$. This simplifies as $\log_2 64 = 3\log_2 4$, or $6 = 3 \cdot 2$.
$\log_2 512 = 3\log_2 8$	Evaluate 8^3.	
$9 = 3 \cdot 3$	Evaluate each of the two logarithms.	
$9 = 9$		

Root property: $\log_b \sqrt[n]{x} = \dfrac{\log_b x}{n}$

$\log_2 \sqrt[3]{8} = \dfrac{\log_2 8}{3}$	Substitute 8 for x, 2 for b, and 3 for n.	Notice that the root property is actually just the power property with a fractional value of n. The cube root is the same as raising to a power of 1/3, so the expression $\log_2\sqrt[3]{8}$ is equivalent to $\log_2 8^{1/3}$. Using the power property, we can rewrite that as $1/3 \log_2 8$, which is the same as $\dfrac{\log_2 8}{3}$.
$\log_2 2 = \dfrac{\log_2 8}{3}$	Evaluate $\sqrt[3]{8}$.	
$1 = 3/3$	Evaluate each of the two logarithms.	
$1 = 1$		

EXAMPLE

Prove the quotient property using other logarithmic identities.

We must prove that $\log_b(x/y) = \log_b x - \log_b y$. The fraction x/y is the same as the product $x \cdot 1/y$, so we can use the product property of logarithms.

$\log_b(x/y) = \log_b(x \cdot 1/y)$	Rewrite x/y as $x \cdot 1/y$.
$= \log_b xy^{-1}$	Rewrite $1/y$ as y^{-1}.
$= \log_b x + \log_b y^{-1}$	Use the logarithm product property.
$= \log_b x + (-1)\log_b y$	Use the logarithm power property.
$= \log_b x - \log_b y$	Addition of a negative is a subtraction.

Another important logarithmic identity involves changing the base.

Change of base property: $\log_b x = \dfrac{\log_k x}{\log_k b}$

EXAMPLE

Use the change of base property to rewrite $\log_{27} 81$ using a base of 3, then simplify.

Try applying the change of base property to the logarithms in Example 4.

$\log_{16} 4 = \dfrac{\log_2 4}{\log_2 16} = 2/4$
$= 1/2$

$\log_9 27 = \dfrac{\log_3 27}{\log_3 9} = 3/2$

$\log_{27} 81 = \dfrac{\log_3 81}{\log_3 27}$	Use the change of base property.
$= 4/3$	Evaluate each logarithm.

Let's use the definition of logarithm to check our answer.

$27^{4/3} \stackrel{?}{=} 81$	Rewrite the equation $\log_{27} 81 = 4/3$ as an exponential relationship.
$(\sqrt[3]{27})^4 \stackrel{?}{=} 81$	Rewrite $27^{4/3}$ as $(27^{1/3})^4$, or $(\sqrt[3]{27})^4$.
$3^4 \stackrel{?}{=} 81$	Evaluate $\sqrt[3]{27}$.
$81 = 81$	Evaluate 3^4.

If possible, choose a base that allows you to directly evaluate each of the logarithms in the fraction, as in Example 12. However, even when this is not possible, the change of base property is still helpful for evaluating logarithms using technology. Any logarithm can be rewritten as a ratio of common logarithms or as a ratio of natural logarithms, each of which can be easily calculated on a calculator, as long as it has a "log" key or an "ln" key.

What is the approximate value of $\log_3 16$?

The numbers 3 and 16 do not share any common factors besides 1, so there is not a convenient new base to use to solve this by hand. Rewrite the given logarithm using the change of base property, as a ratio of common logarithms.

Alternatively, you could use natural logarithms: $\dfrac{\ln 16}{\ln 3}$.

$$\log_3 16 = \frac{\log_{10} 16}{\log_{10} 3}$$

Use your calculator to find the values of $\log 16$ and $\log 3$, and to perform the division.

$\log_3 16 \approx 2.5$, rounded to the nearest tenth.

Even without a calculator, we can approximate logarithms by finding the whole numbers between which a logarithm value lies. Logarithms of a given base are either always increasing or always decreasing as the argument of the logarithm (the x in $\log_b x$) increases. So, if the value of x is between q and r, then the value of $\log_b x$ is between $\log_b q$ and $\log_b r$.

To find which whole numbers $\log_3 16$ lies between, to check our answer to Example 13, find the powers of 3 that 16 lies between. The number 16 is between 3^2, or 9, and 3^3, or 27. Logarithms of base 3 increase for arguments from 9 to 27, so $\log_3 16$ is between $\log_3 9$ and $\log_3 27$, or $\log_3 3^2$ and $\log_3 3^3$. Simplifying these logarithms, we can see that $\log_3 16$ is between 2 and 3. Our answer of 2.5 makes sense.

This is similar to the process you would use to estimate a square root. For example, to find the approximate value of $\sqrt{19}$, you would find the closest two perfect squares to 19. The number 19 is between 16 and 25, so $\sqrt{19}$ is between $\sqrt{16}$ and $\sqrt{25}$, or between 4 and 5.

The various logarithmic identities also come in handy for rewriting equations to solve for a single variable.

Here is how you may see logarithmic identities on the ACT.

When $x > 1$, $5\log_x x^3 = ?$

    A. -15

    B. $-3/5$

    C. $3/5$

    D. $5/3$

    E. 2

EXAMPLE 14

Solve the following equation for *t*.

$$\ln 4^{2t-1} = 8$$

Start by using the power property, because this will pull the expression containing *t* out as a factor, which will make it easier to then isolate *t*.

$(2t - 1) \cdot \ln 4 = 8$	Use the logarithmic power property.
$2t \cdot \ln 4 - \ln 4 = 8$	Distribute the multiplication by ln4 to both terms in the binomial.
$2t \cdot \ln 4 = 8 + \ln 4$	Add ln4 to both sides.
$2t = \dfrac{8 + \ln 4}{\ln 4}$	Divide both sides by ln4.
$t = \dfrac{8 + \ln 4}{2\ln 4}$	Divide both sides by 2.

The answer may also be written as $t = \dfrac{4}{\ln 4} + 1/2$. If we use a calculator, we can solve for the approximate value of *t*, 3.39.

EXAMPLE 15

Solve the following equation for *x*.

$$18 \cdot 10^{4x} = 6$$

The expression log(1/3) is the same as log3⁻¹, which can be rewritten using the power property as −1log3. Our answer to Example 15 can be written as $x = \dfrac{-\log 3}{4}$, which again is equal to about −0.12.

Start by dividing both sides by 18, to isolate the exponential expression containing *x*.

$10^{4x} = 6/18$	
$10^{4x} = 1/3$	Simplify the fraction.
$\log 10^{4x} = \log(1/3)$	Take the common logarithm of both sides.
$4x = \log(1/3)$	Because the base is 10, you can use the identity $\log_b b^n = n$.
$x = \dfrac{\log \dfrac{1}{3}}{4}$	Divide both sides by 4.

Using your calculator, you can evaluate this expression to find that $x \approx -0.12$.

Solve the equation below, for k to the nearest tenth.

$$\log_4(8/k) = 5/4$$

Start by using the logarithmic quotient property to rewrite the left side of the equation.

$\log_4 8 - \log_4 k = 5/4$

$3/2 - \log_4 k = 5/4$ Evaluate $\log_4 8$.

$1/4 - \log_4 k = 0$ Subtract 5/4 from both sides.

$1/4 = \log_4 k$ Add $\log_4 k$ to both sides.

$4^{1/4} = k$ Use the definition of logarithm.

$k \approx 1.4$ Evaluate $4^{1/4}$ on your calculator.

Perhaps you can immediately recognize that $(\sqrt{4})^3 = 8$, or $(4^{1/2})^3 = 8$, so $\log_4 8 = 3/2$. But, if not, you can rewrite $\log_4 8$ as $\log_4 2^3$, and use the logarithmic power property to write it as $3\log_4 2$. The square root (power of 1/2) of 4 is 2, so $\log_4 2 = 1/2$. So, $3\log_4 2 = 3/2$.

$5\log_x x^{-3} = \log_x (x^{-3})^5$ Use the power property of logarithms.

$= \log_x x^{-15}$ Raise x^{-3} to the power of 5: $(x^{-3})^5 = x^{-3 \cdot 5}$.

$= -15$ Use the identity $\log_b b^n = n$.

Alternatively, you could first evaluate $\log_x x^{-3}$ and then multiply by 5.

$5\log_x x^{-3} = 5 \cdot (-3)$ Use the identity $\log_b b^n = n$.

$= -15$ Multiply.

The correct answer is (A).

Lesson 5.4
Logarithmic Functions

REVIEW

For a function $f(x)$:

- The graph of $f(-x)$ is the reflection of $f(x)$ across the y-axis.

- The graph of $-f(x)$ is the reflection of $f(x)$ across the x-axis.

- Multiplying the x-value or the function value by a constant changes the scale of $f(x)$.

In Lesson 5.1, we learned that functions $f(x)$ and $g(x)$ are inverses of one another if $f(g(x)) = x$ and $g(f(x)) = x$. In Lesson 5.2, we learned that $\log_b b^x = x$ and $b^{\log_b x} = x$. If we define $f(x) = b^x$ and $g(x) = \log_b x$, then this demonstrates that $f(x)$ and $g(x)$ are inverse functions.

$$f(g(x)) = f(\log_b x) = b^{\log_b x} = x$$

$$g(f(x)) = g(b^x) = \log_b b^x = x$$

So, $f(x) = b^x$ and $g(x) = \log_b x$ are inverse functions, reflecting how exponentiation and logarithm are inverse operations. The inverse functions for more complicated exponential functions are also logarithmic functions.

EXAMPLE 17

Given that $f(x) = 5 \cdot 7^{2x}$, what is $f^{-1}(x)$?

Replace $f(x)$ with y, switch the x and y variables, and solve for the new y.

$y = 5 \cdot 7^{2x}$	
$x = 5 \cdot 7^{2y}$	Switch x and y.
$x/5 = 7^{2y}$	Divide both sides by 5.
$\log_7(x/5) = 2y$	Take the logarithm, base 7, of both sides.
$\log_7 x - \log_7 5 = 2y$	Use the logarithmic quotient property.
$\dfrac{\log_7 x - \log_7 5}{2} = y$	Divide both sides by 2.

So, $f^{-1}(x) = \dfrac{\log_7 x - \log_7 5}{2}$.

17

What is the inverse function of $p(x) = 2 \cdot 4^{x+9}$?

Use y for $p(x)$, switch the x and y variables, and solve for the new y.

$y = 2 \cdot 4^{x+9}$	
$x = 2 \cdot 4^{y+9}$	Switch x and y.
$x/2 = 4^{y+9}$	Divide both sides by 2.
$\log_4(x/2) = y + 9$	Take the logarithm, base 4, of both sides.
$\log_4 x - \log_4 2 = y + 9$	Use the logarithmic quotient property.
$\log_4 x - 1/2 = y + 9$	Evaluate $\log_4 2$.
$\log_4 x - 19/2 = y$	Subtract 9 from both sides.

So, $p^{-1}(x) = \log_4 x - 19/2$.

Alternatively, after switching x and y, we could have rewritten the right side of the equation to be a single exponential expression of base 2 raised to a power.

$x = 2 \cdot (2^2)^{y+9}$	Rewrite 4 as 2^2.
$x = 2 \cdot 2^{2y+18}$	Raise 2^2 to the power of $(y + 9)$ by multiplying the exponents.
$x = 2^{2y+19}$	Combine the two base-2 exponential terms: $2^1 \cdot 2^{2y+18} = 2^{1+2y+18}$.
$\log_2 x = 2y + 19$	Take the logarithm, base 2, of both sides.
$\log_2 x - 19 = 2y$	Subtract 19 from both sides.
$\dfrac{\log_2 x - 19}{2}$	Divide both sides by 2.

This answer ($p^{-1}(x) = \dfrac{\log_2 x - 19}{2}$ or $p^{-1}(x) = \dfrac{\log_2 x}{2} - 19/2$) would be fine for the given question, but it doesn't look the same as our first answer. Let's use the properties of logarithms to show that they are actually equivalent. We want to convert this equation to our original answer, $y = \log_4 x - 19/2$, which uses a base-4 logarithm.

$y = \dfrac{\log_2 x}{2} - 19/2$

$y = \dfrac{\dfrac{\log_4 x}{\log_4 2}}{2} - 19/2$ Use the change of base property to write $\log_2 x$ as a quotient of base-4 logarithms.

$y = \dfrac{\dfrac{\log_4 x}{\frac{1}{2}}}{2} - 19/2$ Evaluate $\log_4 2$.

$y = \dfrac{2\log_4 x}{2} - 19/2$ Replace division of $\log_4 x$ by $1/2$ with multiplication by its reciprocal, 2.

$y = \log_4 x - 19/2$ Cancel out the common factor of 2.

This produces the same inverse function equation that we originally found, $p^{-1}(x) = \log_4 x - 19/2$.

GRAPHING LOGARITHMIC FUNCTIONS

Logarithmic functions are inverses of exponential functions, so their graphs are related.

As we learned in Lesson 5.1, the graphs of a function and its inverse function are reflections across the line $y = x$. This can help you graph logarithmic functions, as reflections of exponential functions.

EXAMPLE 19

Find the inverse function of $f(x) = 2^x$ and graph both functions on the same coordinate grid. Compare the two graphs.

$y = 2^x$ Replace $f(x)$ with y.

$x = 2^y$ Switch x and y.

$\log_2 x = y$ Take the logarithm, base 2, of both sides.

So, $f^{-1}(x) = \log_2 x$.

Using graphing technology or tables of values, graph $f(x)$ and $f^{-1}(x)$ on the same coordinate grid.

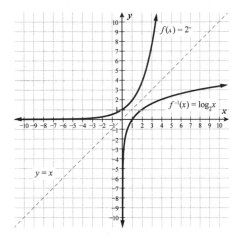

19

You may also recognize that this is an example of an exponential function of the form $f(x) = b^x$, which has an inverse function of $f^{-1}(x) = \log_b x$, with a b-value of 2. So, the function $f(x) = 2^x$ has an inverse function of $f^{-1}(x) = \log_2 x$.

Switching x and y in the given exponential equation replaced each point (a, b) with point (b, a), reflecting the graph of $f(x)$ across the line $y = x$, as we also saw previously in Example 2 for a quadratic function.

These inverse functions are reflections of one another across the line $y = x$. Both graphs are always increasing. The domain of $f(x) = 2^x$ is all real numbers, while its range is all real numbers greater than 0 (there is a horizontal asymptote at the x-axis for its left arm). The domain of $f^{-1}(x)$ is all real numbers greater than 0 (there is a vertical asymptote at the y-axis for its left arm), while its range is all real numbers. This illustrates how the domain of a function is the range of its inverse and the range of a function is the domain of its inverse.

Recall from Lesson 5.2 that $\log_b 0$ and $\log_b c$ for $c < 0$ are undefined. In other words, $y = \log_b x$ is only defined for x-values greater than 0, which is apparent from its graph.

A logarithmic function of the form $y = \log_b x$ has a domain of $x > 0$ and a range of all real numbers.

In Lesson 5.2, you learned that $\log_b 1 = 0$ and $\log_b b = 1$. Because $\log_b 1$ has a value of 0, no matter what value b has, a function of the form $y = \log_b x$ always has an x-intercept of 1. When $x = b$ in a function of this form, $y = \log_b b$, which equals 1. So, $y = \log_b x$ passes through the point $(b, 1)$.

The graph of $f(x) = 2^x$ passes through $(0, 1)$, because $2^0 = 1$, while $f^{-1}(x) = \log_2 x$ passes through $(1, 0)$, because $\log_2 1 = 0$. The graph $f(x) = 2^x$ also passes through $(1, 2)$, while $f^{-1}(x) = \log_2 x$ passes through $(2, 1)$. This illustrates how the inverse of a function transforms each (x, y) point on the original function to (y, x).

> A logarithmic function of the form $y = \log_b x$ has an x-intercept of 1, meaning that it passes through the point $(1, 0)$. A function $y = \log_b x$ also passes through the point $(b, 1)$.

Notice also that the graph of $f^{-1}(x) = \log_2 x$ is below the x-axis for positive values less than 1. This reflects the fact that a base-2 logarithm of a proper fraction must be negative. Remember the meaning of logarithm as an exponent. The base 2 must be raised to a negative exponent to produce an x-value of 1/2, 1/4, or any positive value less than 1.

EXAMPLE

Compare the graphs of $f(x) = \log_3 x$ and $g(x) = \log_{1/3} x$.

We know that the graph of $f(x) = \log_3 x$ passes through $(1, 0)$ and $(3, 1)$, and it has a vertical asymptote of the y-axis for its left end.

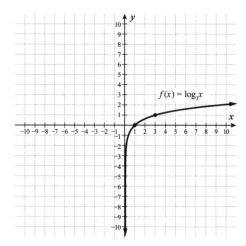

The graph of $g(x) = \log_{1/3} x$ also has a vertical asymptote of the y-axis for its left end, and passes through the points (1, 0) and (1/3, 1). When $x = 3$, $\log_{1/3} x = -1$, so the graph also passes through (3, −1). The graph of $g(x) = \log_{1/3} x$ is shown below.

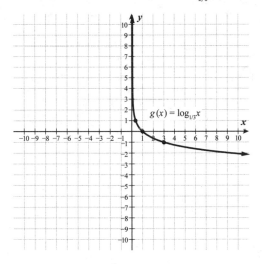

As you can see, this graph is always decreasing, whereas $f(x)$ was always increasing. In fact, the graph of $g(x)$ is the reflection of $f(x)$ across the x-axis. If you rewrite the relationship $g = \log_{1/3} x$ as $(1/3)^g = x$, it is the same as $3^{-g} = x$, or $\log_3 x = -g$. So, $g = -\log_3 x$. In other words, $g(x) = -f(x)$, so the graph of $g(x)$ is a reflection of $f(x)$ across the x-axis.

A logarithmic function of the form $y = \log_b x$ is always increasing if $b > 1$ and is always decreasing if $b < 1$.

As with other types of functions, logarithmic functions are translated vertically when a constant is added to the function value and translated horizontally when a constant is added to x within the function. Their scale, or stretch, is affected if the x-value or the function value is multiplied by a constant.

In Lesson 1.4, we explained how all functions $f(x)$, including polynomial functions, are translated vertically k units for $f(x) + k$ or horizontally h units for $f(x - h)$.

A logarithmic function of the form $y = a \cdot \log_b c(x - h) + v$ is the function $y = \log_b x$ adjusted as follows:

- When $|a| > 1$, the graph is stretched vertically. When $|a| < 1$, the graph is shrunk vertically.

- When $|c| > 1$, the graph is shrunk horizontally. When $|c| < 1$, the graph is stretched horizontally.

- The graph is shifted horizontally h units: to the right if h is positive (when a constant is subtracted from x), to the left if h is negative (when a constant is added to x).

- The graph is shifted vertically v units: up if v is positive, down if v is negative.

- When a is negative, the graph is reflected across the line $y = v$ (the x-axis when $v = 0$).

- When c is negative, the graph is reflected across the line $x = h$ (the y-axis when $h = 0$).

EXAMPLE 21

Graph the function $p(x) = \log_4(x - 3) + 5$. What are the domain and range of $p(x)$?

The graph of $y = \log_4 x$ is always increasing and passes through $(1, 0)$ and $(4, 1)$, as shown below.

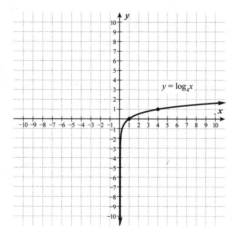

The graph of $p(x) = \log_4(x - 3) + 5$ is the graph of $y = \log_4 x$ shifted 3 units to the right and 5 units up. The vertical asymptote will be moved 3 units to the right, to $x = 3$. Instead of passing through (1, 0) and (4, 1), $p(x)$ will pass through points 3 units right and 5 units up from each of those: (4, 5) and (7, 6).

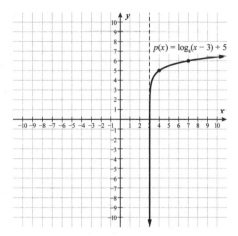

Because the graph is shifted 3 units to the right, the domain is now $x > 3$. This is also apparent from the equation. The argument of the logarithm, $(x - 3)$, must be greater than 0 in order for the logarithm to be defined, so x must be greater than 3. The range, however, is still all real numbers.

No matter how much you shift the parent graph $y = \log_b x$ vertically or adjust its scale, its range will still be all real numbers.

EXAMPLE ㉒

Use technology to graph the function $f(x) = 3e^x$. Algebraically solve for the inverse function, $g(x)$, and graph it.

Here is the graph of $f(x) = 3e^x$.

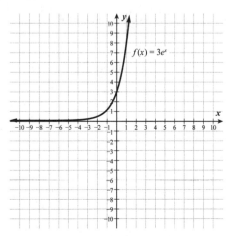

$x = 3e^y$ • Switch x and y in $y = 3e^x$.

$x/3 = e^y$ Divide both sides by 3.

$\ln(x/3) = y$ Take the ln of both sides.

$\ln x - \ln 3 = y$ Use the logarithmic quotient property to rewrite $\ln(x/3)$.

The value of ln3 is a constant approximately equal to 1.1, so the graph of $g(x)$ is the graph of ln$x$ (which is approximately $\log_{2.7} x$) shifted about 1.1 units down. Even without graphing technology, we can get a sense of what this graph looks like.

So, $g(x) = \ln x - \ln 3$. Using technology, we can graph $g(x)$, as shown below. As expected, it is the reflection of $f(x) = 3e^x$ across the line $y = x$.

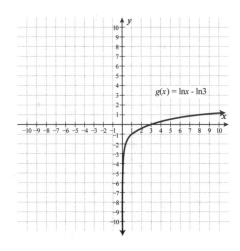

$g(x) = \ln x - \ln 3$

EXAMPLE 23

Graph the function $s(x) = \log_3(-x - 2)$ without technology.

Put this function into the form $y = a \cdot \log_b c(x - h) + v$. In this case, the value of a is 1 and the value of v is 0, so we just need to write the function in the form $y = \log_b c(x - h)$. The x within the parentheses must be positive, as shown, so c must equal -1.

$s(x) = \log_3 -1(x + 2)$ Factor out -1 in the argument.

We can graph $s(x)$ based on the graph of $y = \log_3 x$. The graph of $y = \log_3 x$ is always increasing, passes through (1, 0) and (3, 1), and has a left arm pointing down along a vertical asymptote of the y-axis. It also passes through the point (9, 2), because $\log_3 9 = 2$. Here is the graph of $y = \log_3 x$.

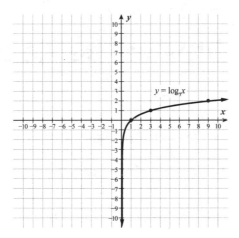

The graph of $s(x) = \log_3 -1(x + 2)$ is the graph $y = \log_3 x$ translated 2 units to the left and reflected across the vertical line $x = -2$, as shown below.

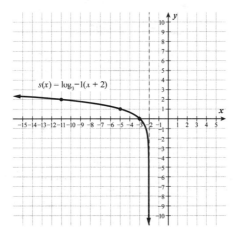

In the form $y = \log_b c(x - h)$, the value of h in $y = \log_3 -1(x + 2)$ is -2. The addition of 2 to x is the same as a subtraction of -2.

Lesson 5.5
Logarithms in the Real World

The ratio $\frac{t}{t_{1/2}}$ is the total amount of elapsed time divided by the amount of time it takes the substance to reduce by half, so this ratio is really just the number of times the substance reduces by half in that time period. So, the original amount, A, is halved (multiplied by 1/2) repeatedly, this number of times, to end up with N, the remaining amount.

Any exponential relationship is, when viewed in the other direction, a logarithmic relationship. So, any real-life situation involving exponential growth (such as growth of certain populations) or decay (such as certain rates of erosion) can be represented by a logarithmic function.

Scientists have studied and described the exponential decay of various substances in terms of their half-life, which is the amount of time it takes half of an original quantity of the substance to decay. The formula $N = A\left(\dfrac{1}{2}\right)^{\frac{t}{t_{1/2}}}$ represents this relationship, where A represents the original quantity of the substance, N represents the quantity of the substance left after an amount of time, t, and $t_{1/2}$ represents the half-life of that particular substance.

EXAMPLE 24

Iodine-131 has a half-life of about 8 days. Write an equation that gives the amount of time, in days, elapsed since an iodine-131 sample weighed 10 milligrams, given its current weight in milligrams. Using technology, graph both the exponential function and the logarithmic function that represent this relationship. Solve for the time it takes the 10-milligram sample to reduce to 1.25 milligrams.

Notice that time is defined in seconds in this formula, unlike the first equation we will write, which uses days.

The relationship between the original quantity of an iodine-131 sample, A, and its quantity, N, after t seconds is also given by the formula $N = A \cdot e^{-10^{-6} \cdot t}$. Rewrite this formula to solve for t as a function of N, then show that the resulting function produces the same answer for the amount of time it takes a 10-milligram sample to reduce to 1.25 milligrams.

We'll use the half-life formula, $N = A\left(\dfrac{1}{2}\right)^{\frac{t}{t_{1/2}}}$, for the first equation, with time t expressed in days. Iodine-131 has a half-life of about 8 days, so $t_{1/2} = 8$. The original amount of the sample is 10 milligrams, so $A = 10$. Our exponential equation is therefore $N = 10 \cdot (1/2)^{t/8}$. Let's solve this equation for t in terms of N.

$N/10 = (1/2)^{t/8}$	Divide both sides by 10.
$N/10 = 2^{-t/8}$	Rewrite 1/2 as 2^{-1} and raise it to the power of $t/8$ by multiplying the exponents.
$\log_2(N/10) = -t/8$	Take the logarithm, base 2, of both sides.
$-8\log_2(N/10) = t$	Multiply both sides by −8.

The graphs of $N = 10 \cdot (1/2)^{t/8}$ and $t = -8\log_2(N/10)$ are shown below.

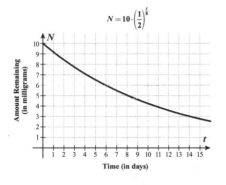

$$N = 10 \cdot \left(\frac{1}{2}\right)^{\frac{t}{8}}$$

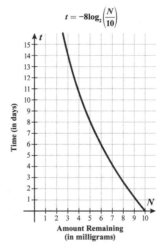

$$t = -8\log_2\left(\frac{N}{10}\right)$$

The exponential function graph may seem more intuitively logical, because time increases along the horizontal axis from left to right, but the logarithmic function graph also accurately represents the relationship. You can see from the graph that, for a quantity, N, of 10 milligrams, the time elapsed is 0 days, and for a quantity, N, of 5 milligrams, the time elapsed is 8 days. This fits with the fact that iodine-131 has a half-life of 8 days; half of the original sample size is left after 8 days.

To solve for the time it takes the 10-milligram sample to decay to a remaining mass of 1.25 milligrams, substitute 1.25 for N in $t = -8\log_2(N/10)$.

$$t = -8\log_2(1.25/10)$$

When using specific variables to represent certain quantities, we solve for the inverse function by solving for the other variable. We do not switch variables the way we do with x and y.

This logarithmic function can also be written as
$t = -8(\log_2 N - \log_2 10)$
using the logarithmic quotient property and then simplified as
$t = -8\log_2 N + 8\log_2 10$.

Notice that, as with any pair of inverse functions, these two graphs are reflections of one another across the diagonal line $y = x$. Any point (t, N) in the graph on the left has a corresponding point (N, t) in the graph on the right.

We could expand either graph to see what t-value corresponds to an N-value of 1.25. Zoom in on a graphing calculator for a pretty accurate answer.

The decimal 0.125 is equal to 1/8, or 8^{-1}, which is the same as 2^{-3}. So, $\log_2 0.125 = \log_2 2^{-3} = -3$.

$t = -8\log_2 0.125$ Divide 1.25 by 10.

$t = -8 \cdot (-3)$ Evaluate $\log_2 0.125$.

$t = 24$

There will be 1.25 milligrams left of the sample after 24 days.

10^{-6} is 0.000001, so the formula can also be written as $N = 10e^{-0.000001t}$.

The second formula is $N = A \cdot e^{-10^{-6} \cdot t}$, where t is time elapsed, in seconds. Again, $A = 10$, so the equation is $N = 10 \cdot e^{-10^{-6} \cdot t}$. Let's solve this equation for t.

The exponent $10^{-6} \cdot t$, or $0.000001t$, is the opposite, of $-10^{-6} \cdot t$, or $-0.000001t$. The -6 in the exponent of 10 only influences the magnitude of the coefficient of t, so it is not affected when you rewrite $10^{-6} \cdot 10^6 = 10^{-6+6} = 10^0 = 1$, so $(10^{-6} \cdot t) \cdot 10^6 = t$.

$N = \dfrac{10}{e^{10^{-6} \cdot t}}$ Rewrite e to a negative exponent as 1 over e raised to a positive version of that exponent.

$N e^{10^{-6} \cdot t} = 10$ Multiply both sides by $e^{10^{-6} \cdot t}$.

$e^{10^{-6} \cdot t} = 10/N$ Divide both sides by N.

$10^{-6} \cdot t = \ln(10/N)$ Take the ln of both sides.

$t = 10^6 \cdot \ln(10/N)$ Multiply both sides by 10^6.

To solve for the time it takes the sample to decay to just 1.25 milligrams, substitute 1.25 for N in this new formula and solve for t.

$t = 10^6 \cdot \ln(10/1.25)$ Divide 10 by 1.25.

$t = 10^6 \cdot \ln 8$ Use your calculator to compute
$t \approx 2{,}079{,}441$ $10^6 \cdot \ln 8$.

The 10-milligram sample decays to 1.25 milligrams in about 2,079,441 seconds. Let's convert seconds to days, to compare this answer with the one we got using the first formula.

$$2{,}079{,}441 \text{ sec} \times \frac{1 \text{ min}}{60 \text{ sec}} \times \frac{1 \text{ hr}}{60 \text{ min}} \times \frac{1 \text{ day}}{24 \text{ hr}} \approx 24$$

In other words, we must divide 2,079,441 by 60 to get the number of minutes, divide again by 60 to get the number of hours, and divide by 24 to finally find the equivalence in days.

This is the same answer (24 days) that we got using the formula $t = -8\log_2(N/10)$.

CHAPTER 5 PRACTICE QUESTIONS

Directions: Complete the following open-ended problems as specified by each question stem. For extra practice after answering each question, try using an alternative method to solve the problem or check your work.

1. Evaluate each of the following without a calculator:

   (a) $\log_5 125$

   (b) $\log_8 64$

   (c) $\log_3 \dfrac{1}{9}$

   (d) $\log_9 9$

   (e) $\log \dfrac{1}{10}$

2. For each of the following, find the inverse function:

   (a) $y = \sqrt{3x - 5}$

   (b) $f(x) = 16 - x^2, x \geq 0$

   (c) $y = \sqrt{9 - x^2}, 0 \leq x \leq 3$

   (d) $g(x) = 3 \cdot 8^{-x}$

   (e) $h(x) = 4 \cdot 6^{2x} - 8$

3. For each of the following, rewrite using properties of logarithms to express as a sum or difference:

   (a) $\log_2 \dfrac{8x^2\sqrt{z}}{y^3}$

   (b) $\ln \dfrac{\sqrt[3]{4a - 5}}{3b}$

   (c) $\ln \sqrt{x^2\left(x^2 + 2\right)}$

   (d) $\log_z \dfrac{\sqrt{ac^{\,5}}}{b^6}$

4. Graph the function $f(x) = \log_{1/2}(x + 7) - 3$.

5. For each of the following, solve for x without a calculator:

   (a) $9^{x+2} = 27^2$

   (b) $\log_2 \dfrac{1}{16} = x^2 - 4x$

   (c) $\log_2(x^2 + 12) = \log_2 7x$

   (d) $2\log_5 x = \log_5 9, x > 0$

6. For each of the following, use the properties of logarithms to rewrite the given expressions as a single logarithm:

   (a) $\log_5(x^2 - 1) - 6\log_5(x + 1)$

   (b) $\log\left(\dfrac{x^2 - 2x - 15}{x^2 - 5x}\right) - \log\left(\dfrac{x^2 + 6x + 9}{x}\right)$

   (c) $6\log_3\sqrt{4x - 3} - \log_3(\dfrac{2}{x}) + \log_3 9$

   (d) $3\log_2\sqrt[3]{a} - c\log_2 b + 2\log_2(de)$

7. For each of the following, solve for x. Make sure that your solutions do not make any terms in the original equations undefined.

   (a) $4^{x+3} = 7^x$

   (b) $\ln(x + 5) = \ln(x - 1) - \ln(x + 1)$

   (c) $\log_{10}(x^2 + 5x) - \log_{10} x = 2$

   (d) $\ln(4x - 3) = \ln 25$

8. A population of fungi grows exponentially. A population that initially has 10,000 grows to 25,000 after 2 hours.

   (a) Use the exponential growth function, $A(t) = A_0 e^{kt}$ to find the value of k rounded to 2 decimal places. What will be the fungus population after 6 hours?

   (b) Write a logarithmic function that gives t as a function of $A(t)$, which we can now write as A.

   (c) If a scientist estimates that the fungus population in this sample is 63,000, then how many hours have passed since the initial measure of 10,000 individuals in the population?

9. A certain species of marmoset is listed as vulnerable for its conservation status. There are currently about 8000 of this species, which is 46% of the population 6 years ago, and the decrease in population has been exponential. Let t represent the number of years since 6 years ago (so $t = 6$ represents the current year).

   (a) Write an exponential function that gives the population of this marmoset species in a given year, as defined by t. (Use the formula for continuous growth/decay, $A = A_0 e^{kt}$, where A is the population after t years, given that the initial population, at $t = 0$, is A_0.)

   (b) Rewrite the equation to give t as a function of the population.

   (c) If the population drops below 2500, the species will be considered endangered. If the population continues to decrease at the same exponential rate, in how many years will this marmoset species be considered endangered?

SOLUTIONS TO CHAPTER 5 PRACTICE QUESTIONS

1. **(a) 3, (b) 2, (c) –2, (d) 1, (e) –1**

   (a) Recall that $\log_5 125$ is the same as saying "5 raised to what power is equal to 125." 5 raised to the third power equals 125, so $\log_5 125$ equals 3.

   (b) Recall that $\log_8 64$ is the same as saying "8 raised to what power is equal to 64." 8 squared equals 64, so $\log_8 64$ equals 2.

   (c) Recall that $\log_3 \dfrac{1}{9}$ is the same as saying "3 raised to what power is equal to 1/9." 3 raised to the power of –2 equals 1/9, so $\log_3 1/9$ equals –2.

   (d) Recall that $\log_9 9$ is the same as saying "9 raised to what power is equal to 9." 9 raised to the first power equals 9, so $\log_9 9$ equals 1.

   (e) Recall that a logarithm without a provided base has a base of 10. $\log \dfrac{1}{10}$ is the same as saying "10 raised to what power is equal to 1/10." 10 raised to the –1 power is 1/10, so $\log \dfrac{1}{10}$ equals –1.

2. **(a)** $(x^2 + 5)/3 = y, x \geq 0$; **(b)** $f^{-1}(x) = \sqrt{16 - x}$; **(c)** $y = \sqrt{9 - x^2}, 0 \leq x \leq 3$; **(d)** $g^{-1}(x) = -\log_8(x/3)$; **(e)** $h^{-1}(x) = (\log_6(x/4) + 2)/2$

   (a)

$y = \sqrt{3x - 5}$	Given.
$x = \sqrt{3y - 5}$	Interchange x and y.
$x^2 = 3y - 5$	Square both sides.
$x^2 + 5 = 3y$	Add 5 to both sides.
$\dfrac{x^2 + 5}{3} = y, x \geq 0$	Divide both sides by 3, and note domain restriction based on the range restriction of the given function.

   (b)

$f(x) = 16 - x^2, x \geq 0$	Given.
$y = 16 - x^2$	Replace $f(x)$ with y.
$x = 16 - y^2, y \geq 0$	Interchange x and y; given domain is range of inverse.
$y^2 = 16 - x$	Add y^2 to and subtract x from both sides.
$y = \pm \sqrt{16 - x}$	Take square root of both sides.
$y = \sqrt{16 - x}$	Apply restriction.
$f^{-1}(x) = \sqrt{16 - x}$	Replace y with $f^{-1}(x)$.

DRILL

(c) $y = \sqrt{9 - x^2}, 0 \le x \le 3$ — Given.

$x = \sqrt{9 - y^2}, 0 \le y \le 3$ — Interchange x and y; given domain is range of inverse.

$x^2 = 9 - y^2$ — Square both sides.

$y^2 = 9 - x^2$ — Add y^2 to and subtract x^2 from both sides.

$y = \pm\sqrt{9 - x^2}$ — Take square root of both sides.

$y = \sqrt{9 - x^2}, 0 \le x \le 3$ — Apply restriction to the range, and note domain restriction based on the range restriction of the original function.

(d) $y = 3 \cdot 8^{-x}$ — Replace $g(x)$ with y.

$x = 3 \cdot 8^{-y}$ — Interchange x and y.

$x/3 = 8^{-y}$ — Divide both sides by 3.

$\log_8(x/3) = -y$ — Take the logarithm, base 8, of both sides.

$-\log_8(x/3) = y$ — Multiply both sides by −1.

$g^{-1}(x) = -\log_8(x/3)$ — Replace y with $g^{-1}(x)$.

We can also use the logarithmic quotient property to write this inverse function as $g^{-1}(x) = -(\log_8 x - \log_8 3)$, which simplifies as $g^{-1}(x) = -\log_8 x + \log_8 3$.

(e) $y = 4 \cdot 6^{2x} - 8$ — Replace $h(x)$ with y.

$x = 4 \cdot 6^{2y} - 8$ — Interchange x and y.

$x + 8 = 4 \cdot 6^{2y}$ — Add 8 to both sides.

$x/4 + 2 = 6^{2y}$ — Divide both sides by 4.

$\log_6(x/4 + 2) = 2y$ — Take the logarithm, base 6, of both sides.

$$\frac{\log_6\left(\dfrac{x}{4} + 2\right)}{2} = y$$ — Divide both sides by 2.

$$h^{-1}(x) = \frac{\log_6\left(\dfrac{x}{4} + 2\right)}{2}$$ — Replace y with $h^{-1}(x)$.

We can also write this inverse function as $h^{-1}(x) = \log_6\sqrt{\dfrac{x}{4} + 2}$, using the logarithmic root property.

3.   See explanations.

(a) $\log_2 \dfrac{8\,x^2\,\sqrt{z}}{y^3}$

Given.

$\log_2 \dfrac{8\,x^2\,(z)^{\frac{1}{2}}}{y^3}$

Rewrite z term with an exponent.

$\log_2 8 + \log_2 x^2 + \log_2(z)^{\frac{1}{2}} - \log_2 y^3$

Rewrite using product and quotient properties.

$\log_2 8 + 2\log_2 x + \dfrac{1}{2}\log_2 z - 3\log_2 y$

Rewrite using power property.

$3 + 2\log_2 x + \dfrac{1}{2}\log_2 z - 3\log_2 y$

Evaluate the first term.

(b) $\ln \dfrac{\sqrt[3]{4a-5}}{3b}$

Given.

$\ln \dfrac{(4a-5)^{\frac{1}{3}}}{3b}$

Rewrite numerator with exponent.

$\ln(4a-5)^{\frac{1}{3}} - \ln 3b$

Rewrite using quotient property.

$\dfrac{1}{3}\ln(4a-5) - \ln 3b$

Rewrite first term using power property.

(c) $\ln \sqrt{x^2\left(x^2+2\right)}$

Given.

$\ln\left[x^2\left(x^2+2\right)\right]^{\frac{1}{2}}$

Rewrite term with an exponent.

$\dfrac{1}{2}\ln\left[x^2\left(x^2+2\right)\right]$

Rewrite using power property.

$\dfrac{1}{2}[\ln x^2 + \ln(x^2+2)]$

Rewrite using product property.

$\dfrac{1}{2}[2\ln x + \ln(x^2+2)]$

Rewrite first term using power property.

$\dfrac{1}{2}(2\ln x) + \dfrac{1}{2}\ln(x^2+2)$

Distribute $\dfrac{1}{2}$ to each term.

$\ln x + \dfrac{1}{2}\ln(x^2+2)$

Simplify first term.

DRILL

(d) $\log_z \dfrac{\sqrt{a}c^5}{b^6}$

Given.

$\log_z \dfrac{(a)^{\frac{1}{2}}c^5}{b^6}$

Rewrite terms with exponents.

$\log_z(a)^{\frac{1}{2}} + \log_z c^5 - \log_z b^6$

Rewrite using product and quotient properties.

$\dfrac{1}{2}\log_z a + 5\log_z c - 6\log_z b$

Rewrite using power property.

4. **See graph.**
   The value of the base, 1/2, is less than 1, so this function is always decreasing.

   The function $y = \log_{1/2}x$ has a vertical asymptote of the y-axis for its left end, because an x-value of 0 makes it undefined, and it passes through the points (1, 0), (2, –1), (4, –2), and (8, –3), because $\log_{1/2}1 = 0$, $\log_{1/2}2 = -1$, $\log_{1/2}4 = -2$, and $\log_{1/2}8 = -3$. The graph of $f(x) = \log_{1/2}(x + 7) - 3$ is the graph of $y = \log_{1/2}x$ shifted 7 units to the left and 3 units down. So, the graph of $f(x)$ has a vertical asymptote of $x = -7$ and passes through the points with x-coordinates 7 less and y-coordinates 3 less than those of points on $y = \log_{1/2}x$.

   $(1 - 7, 0 - 3) \rightarrow (-6, -3)$
   $(2 - 7, -1 - 3) \rightarrow (-5, -4)$
   $(4 - 7, -2 - 3) \rightarrow (-3, -5)$
   $(8 - 7, -3 - 3) \rightarrow (1, -6)$

   The graph of $f(x) = \log_{1/2}(x + 7) - 3$ passes through (–6, –3), (–5, –4), (–3, –5), and (1, –6). The graph of $f(x)$ is shown below, with asymptote indicated by a dashed line.

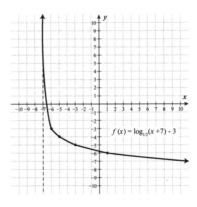

$f(x) = \log_{1/2}(x + 7) - 3$

5. **(a) 1, (b) 2, (c) 3, 4 (d) 3**

(a) $9^{x+2} = 27^2$ Given.

$(3^2)^{x+2} = (3^3)^2$ Rewrite using common bases.

$3^{2x+4} = 3^6$ Multiply the exponents on both sides.

$2x + 4 = 6$ Same bases equates the exponents.

$2x = 2$ Subtract 4 from both sides.

$x = 1$ Divide both sides by 2.

(b) $\log_2 \dfrac{1}{16} = x^2 - 4x$ Given.

$x^2 - 4x = -4$ Evaluate $\log_2 1/16$, which is the same as $\log_2(2^{-4})$.

$x^2 - 4x + 4 = 0$ Add 4 to both sides to set the quadratic equal to zero.

$(x - 2)^2 = 0$ Factor the quadratic as the square of the difference.

$x - 2 = 0$ Take a square root of both sides.

$x = 2$ Add 2 to both sides.

(c) $\log_2(x^2 + 12) = \log_2 7x$ Given.

$2^{\log_2 7x} = x^2 + 12$ Rewrite in exponential form.

$7x = x^2 + 12$ Rewrite the left side using $b^{\log_b n} = n$.

$0 = x^2 - 7x + 12$ Subtract 7x from both sides to set the quadratic equal to zero.

$0 = (x - 3)(x - 4)$ Factor the quadratic.

$0 = x - 3; 0 = x - 4$ Set each factor equal to zero.

$3 = x; 4 = x$ Add 3 and 4 to each side to solve for x.

(d) $2 \log_5 x = \log_5 9$ Given.

$\log_5 x^2 = \log_5 9$ Rewrite the left using power property.

$5^{\log_5 9} = x^2$ Rewrite in exponential form.

$9 = x^2$ Rewrite the left side using $b^{\log_b n} = n$.

$\pm 3 = x$ Take a square root of both sides.

$3 = x$ Apply restriction ($x > 0$).

6. **See the following logarithms.**

   (a) $\log_5(x^2 - 1) - 6\log_5(x + 1)$ Given.

        $\log_5(x^2 - 1) - \log_5(x + 1)^6$ Rewrite 2nd term using power property.

        $\log_5\dfrac{\left(x^2 - 1\right)}{\left(x + 1\right)^6}$ Rewrite using quotient property.

        $\log_5\dfrac{\left(x + 1\right)\left(x - 1\right)}{\left(x + 1\right)^6}$ Factor the numerator.

        $\log_5\dfrac{x - 1}{\left(x + 1\right)^5}$ Cancel common factors.

   (b) $\log\left(\dfrac{x^2 - 2x - 15}{x^2 - 5x}\right) - \log\left(\dfrac{x^2 + 6x + 9}{x}\right)$ Given.

        $\log\left(\dfrac{\left(x - 5\right)\left(x + 3\right)}{x\left(x - 5\right)}\right) - \log\left(\dfrac{\left(x + 3\right)\left(x + 3\right)}{x}\right)$ Factor each expression.

        $\log\dfrac{x + 3}{x} - \log\dfrac{\left(x + 3\right)^2}{x}$ Cancel out common factors.

        $\log\dfrac{\dfrac{x + 3}{x}}{\dfrac{\left(x + 3\right)^2}{x}}$ Rewrite using quotient property.

        $\log\left(\dfrac{x + 3}{x} \times \dfrac{x}{\left(x + 3\right)^2}\right)$ Division by fraction equates multiplication by the reciprocal.

        $\log\dfrac{1}{x + 3}$ Cancel out common terms.

        $\log(x + 3)^{-1}$ Rewrite using a negative exponent.

        $-\log(x + 3)$ Rewrite using power property.

   (c) $6\log_3\sqrt{4x - 3} - \log_3(\dfrac{2}{x}) + \log_3 9$ Given.

        $6\log_3(4x - 3)^{\frac{1}{2}} - \log_3\left(\dfrac{x}{2}\right)^{-1} + \log_3(3)^2$ Rewrite terms with exponents.

        $\log_3(4x - 3)^3 + \log_3\left(\dfrac{x}{2}\right)^{-1} + 2$ Rewrite using power property; evaluate final log.

        $\log_3\dfrac{x\left(4x - 3\right)^3}{2}$ Combine using product property.

(d) $3\log_2 \sqrt[3]{a} - c\log_2 b + 2\log_2(de)$ Given.

$3\log_2 a^{\frac{1}{3}} - \log_2 b^c + \log_2(de)^2$ Rewrite terms with exponents.

$\log_2 a - \log_2 b^c + \log_2(de)^2$ Rewrite 1st term using power property.

$\log_2 \dfrac{a(de)^2}{b^c}$ Rewrite using product and quotient properties.

---

7. **(a) 3ln4 / (ln7 − ln4); (b) no solution; (c) 95, (d) 7**

(a) $4^{x+3} = 7^x$ Given.

$\ln(4)^{x+3} = \ln 7^x$ Take ln of both sides.

$(x+3)\ln 4 = x\ln 7$ Rewrite using power property.

$x\ln 4 + 3\ln 4 = x\ln 7$ Distribute the left side.

$3\ln 4 = x\ln 7 - x\ln 4$ Subtract x ln 4 from both sides.

$3\ln 4 = x(\ln 7 - \ln 4)$ Factor x from both terms.

$\dfrac{3\ln 4}{\ln 7 - \ln 4} = x$ Divide both sides by ln 7 − ln4.

(b) $\ln(x+5) = \ln(x-1) - \ln(x+1)$ Given.

$\ln(x+5) = \ln \dfrac{x-1}{x+1}$ Rewrite right side using quotient property.

$x+5 = \dfrac{x-1}{x+1}$ Two logarithms of the same base set equal to one another have equal arguments.

$(x+5)(x+1) = x-1$ Multiply both sides by $(x+1)$.

$x^2 + 6x + 5 = x - 1$ Use FOIL on the left side.

$x^2 + 5x + 6 = 0$ Subtract $(x-1)$ from both sides.

$x = \dfrac{-(5) \pm \sqrt{(5)^2 - 4(1)(6)}}{2(1)}$ Use Quadratic Formula with $a = 1$, $b = 5$, $c = 6$.

$x = \dfrac{-5 \pm \sqrt{1}}{2}$ Simplify radicand.

$x = -2, -3$ Perform arithmetic.

However, the argument of a logarithm cannot be negative, so each of these values make two of the logarithmic expressions in the original equation undefined. There are **no** solutions to the given equation.

(c) $\log_{10}(x^2 + 5x) - \log_{10} x = 2$ Given.

$\log_{10}\dfrac{x^2 + 5x}{x} = 2$ Rewrite using quotient property.

$10^2 = \dfrac{x^2 + 5x}{x}$ Rewrite in exponential form.

$100x = x^2 + 5x$ Square 10; multiply both sides by x.

$0 = x^2 - 95x$ Subtract 100x from both sides.

$0 = x(x - 95)$ Factor x from both terms.

$0 = x \; ; \;\; 0 = x - 95$ Set each term equal to zero and solve.

$95 = x$

An x-value of 0 would make $\log_{10} x$ undefined in the original equation, so that is an extraneous solution. The only solution to the given equation is $x = 95$.

(d) $\ln(4x - 3) = \ln 25$ Given.

$4x - 3 = 25$ Two logarithms of the same base set equal to one another have equal arguments.

$4x = 28$ Add 3 to both sides.

$x = 7$ Divide by 4 on both sides.

8.   **(a) 157,998; (b) $t = 2.17\ln(A/10{,}000)$; (c) 4**

(a)   The question provides the function, the initial value ($t = 0$) of 10,000 (A_o), and the amount after $t = 2$ hours. Plug in the given values to arrive at the value of k:

$A(t) = A_o e^{kt}$ Given.

$A(2) = (10{,}000)e^{2k}$ Let $t = 2$.

$25{,}000 = 10{,}000e^{2k}$ Substitute 25,000 for $A(2)$.

$2.5 = e^{2k}$ Divide both sides by 10,000.

$\ln 2.5 = 2k$ Take ln of both sides.

$\dfrac{\ln 2.5}{2} = k$ Divide both sides by 2.

$0.46 \approx k$ Perform indicated division and round.

Next, with the found value of k rounded to 2 decimal places, evaluate using the function when $t = 6$ hours:

$A(t) = A_o e^{kt}$ Given.

$A(t) = 10{,}000e^{0.46t}$ Substitute original value and value of k.

$A(6) = 10{,}000e^{(0.46)(6)}$ Substitute $t = 6$.

$A(6) \approx 157{,}998$ Use your calculator to evaluate and round off.

(b) We must solve the exponential equation $A(t) = 10,000e^{0.46t}$ for t.

$A = 10,000e^{0.46t}$	Rewrite using A instead of $A(t)$.
$A/10,000 = e^{0.46t}$	Divide both sides by 10,000.
$\ln(A/10,000) = 0.46t$	Take the ln of both sides.
$(1/0.46)\ln(A/10,000) = t$	Divide both sides by 0.46.
$2.17\ln(A/10,000) = t$	Evaluate 1/0.46.

The function that gives t, time in hours passed, as a function of A, the current fungus population in the sample, is $t = 2.17 \ln (A/10,000)$.

(c) If a scientist estimates the population in the sample to be 63,000, then $A = 63,000$. Substitute this value and solve for t.

$t = 2.17\ln(63,000/10,000)$	Substitute 63,000 for A.
$t = 2.17\ln6.3$	Evaluate 63,000/10,000.
$t \approx 4$	Use a calculator to find ln6.3 and multiply that by 2.17.

If the fungus population in the sample is 63,000, then 4 hours have passed since the initial population count of 10,000.

9. **(a) $A = 17,391e^{-0.129t}$; (b) $t = -7.75\ln(A/17,391)$; (c) 9**

(a) If the current population of this marmoset species, A, is 46% of the population 6 years ago, A_0, then A/A_0 is equal to 46%, or 0.46. Solve the equation $A = A_0e^{kt}$ for A/A_0 then substitute the appropriate values.

$A/A_0 = e^{kt}$	Divide both sides of the formula by A_0.
$0.46 = e^{6k}$	Substitute 0.46 for A/A_0 and 6 for t.
$\ln0.46 = 6k$	Take the ln of both sides.
$\dfrac{\ln0.46}{6} = k$	Divide both sides by 6.
$k \approx -0.129$	Evaluate $\dfrac{\ln0.46}{6}$ using a calculator.

By using the information about the population 6 years ago and this year, we have found the exponential decay rate. Substituting back into the formula, we have $A = A_0e^{-0.129t}$. To finish creating an exponential function of A in terms of t, we must find A_0, the population 6 years ago. The current population, 8000, is 46% of A_0.

$$8000 = 0.46A_0$$
$$A_0 \approx 17,391 \qquad \text{Divide both sides by 0.46.}$$

So, the population, A, of the marmoset species t years after 6 years ago is given by the function $A = 17,391e^{-0.129t}$.

(b) Solve the equation found in part (a) for t to rewrite the formula.

$A/17{,}391 = e^{-0.129t}$	Divide both sides by 17,391.
$\ln(A/17{,}391) = -0.129t$	Take the ln of both sides.
$-7.75\ln(A/17{,}391) = t$	Divide both sides by −0.129.

A function for t in terms of A is $t = -7.75\ln(A/17{,}391)$.

(c) To find when the population reaches 2500, substitute 2500 for A and solve for t.

$t = -7.75\ln(2500/17{,}391)$

$t \approx 15$	Use your calculator to evaluate.

The marmoset species population will drop below 2500 about 15 years after 6 years ago, which is 9 years from now.

Alternatively, you could redefine $t = 0$ as the current year, with $A_0 = 8000$ and t representing years from now, in the exponential function.

$A = 8000e^{-0.129t}$

$A/8000 = e^{-0.129t}$	Divide both sides by 8000.
$\ln(A/8000) = -0.129t$	Take the ln of both sides.
$-7.75\ln(A/8000) = t$	Divide both sides by −0.129.
$-7.75\ln(2500/8000) = t$	Substitute 2500 for A.
$t \approx 9$	Use your calculator to evaluate.

This solution also tells you that the marmoset species will be considered endangered about 9 years from now.

REFLECT

**Congratulations on completing Chapter 5!
Here's what we just covered.
Rate your confidence in your ability to**

- Find the inverse relation of a given function, including exponential functions, and determine any necessary domain restrictions on the function to make its inverse a function
  ① ② ③ ④ ⑤

- Evaluate or estimate numerical logarithms, where they exist
  ① ② ③ ④ ⑤

- Solve single-variable equations containing logarithmic expressions
  ① ② ③ ④ ⑤

- Use binary, natural, and common logarithms
  ① ② ③ ④ ⑤

- Use logarithmic identities
  ① ② ③ ④ ⑤

- Graph logarithmic functions
  ① ② ③ ④ ⑤

- Write and use logarithmic functions to represent real-life situations and solve problems
  ① ② ③ ④ ⑤

If you rated any of these topics lower than you'd like, consider reviewing the corresponding lesson before moving on, especially if you found yourself unable to correctly answer one of the related end-of-chapter questions.

Access your online student tools for a handy, printable list of Key Points for this chapter. These can be helpful for retaining what you've learned as you continue to explore these topics.

Chapter 6
More Functions

GOALS By the end of this chapter, you will be able to

- Graph cube root functions

- Graph piecewise-defined functions, including step functions (such as floor and ceiling functions) and absolute value functions

- Solve single-variable absolute value equations, both algebraically and by graphing systems of equations

- Graph exponential functions

- Write and use cube root, piecewise-defined, and exponential functions to represent real-life situations and solve problems

Lesson 6.1
Cube Root Functions

REVIEW

For any function $f(x)$, the graph of $f(x + k)$ is a horizontal translation of $f(x)$ by k units (to the left if k is positive and to the right if k is negative), and $f(x) + k$ is a vertical translation of $f(x)$ by k units (upward if k is positive, downward if k is negative), where k is a constant.

For any function $f(x)$, the graph of $-f(x)$ is a reflection of $f(x)$ across the x-axis, and the graph of $f(-x)$ is a reflection of $f(x)$ across the y-axis.

When a graph is symmetrical with respect to a point, each point in the graph is the same distance from the center point as a point directly opposite it (on a line passing through the center point). If the image is rotated 180° around the center point, it exactly matches the original image.

Cube root functions are, like square root functions, another type of radical function. They are the inverse of cubic functions (sometimes requiring a domain restriction).

The cubic function $y = x^3 - 2$ is shown on the coordinate grid below.

This is similar to what we saw in Example 16 in Lesson 3.6, where we found a square root function as the inverse of a quadratic function (with a domain restriction).

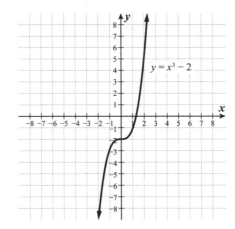

$y = x^3 - 2$

To find the inverse relationship, switch the x and y variables, then solve for the new y.

$x = y^3 - 2$ Switch the x and y in $y = x^3 - 2$.

$x + 2 = y^3$ Add 2 to both sides.

$\sqrt[3]{x + 2} = y$ Take the cube root of both sides.

See Lesson 5.1 for a review of inverse functions, including how to algebraically find the inverse relation for a given function, and how the graphs of inverse functions are related.

The inverse function of $y = x^3 - 2$ is the cube root function $y = \sqrt[3]{x + 2}$. Inverse functions are reflections of one another across the diagonal line $y = x$, so the graph of $y = \sqrt[3]{x + 2}$ is the reflection of $y = x^3 - 2$ across the line $y = x$.

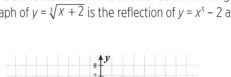
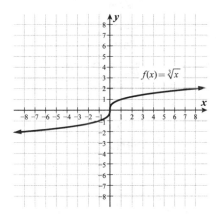

$y = \sqrt[3]{x + 2}$

The parent cube root function, $f(x) = \sqrt[3]{x}$, is the inverse of the parent cubic function, $f(x) = x^3$. The graph of $f(x) = \sqrt[3]{x}$ is shown below.

The sign (positive or negative) is preserved when you cube a number, so the cube root will automatically give the correct sign. This means that we do not need to use the ± symbol that is necessary when taking the square root of both sides of an equation of the form $y^2 = a$.

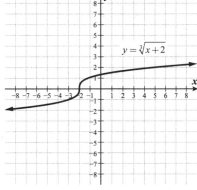

$f(x) = \sqrt[3]{x}$

Comparing the previous graph to this one, you can see that $y = \sqrt[3]{x + 2}$ is the function $f(x) = \sqrt[3]{x}$ translated 2 units to the left.

Graph the function $g(x) = \sqrt[3]{x - 8} + 1$. What are its x- and y-intercepts?

The function $g(x) = \sqrt[3]{x - 8} + 1$ is the function $f(x) = \sqrt[3]{x}$ translated 8 units to the right and 1 unit up.

We could also create a table of values to find points on the graph. The value of $g(0) = -1$, the value of $g(7) = 0$, the value of $g(8) = 1$, and the value of $g(9) = 2$.

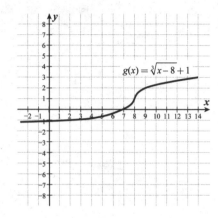

$g(x) = \sqrt[3]{x - 8} + 1$

If we have graphed accurately or used graphing technology, we can visually identify the x- and y-intercepts. The graph has an x-intercept of 7 and a y-intercept of −1. Let's use the function equation to confirm that these intercepts are correct.

The x-intercept occurs when $y = 0$, or $g(x) = 0$.

$0 = \sqrt[3]{x - 8} + 1$	Substitute 0 for $g(x)$ in the function equation.
$-1 = \sqrt[3]{x - 8}$	Subtract 1 from both sides, to isolate the cube root.
$(-1)^3 = x - 8$	Cube both sides.
$-1 = x - 8$	Evaluate $(-1)^3$.
$7 = x$	Add 8 to both sides.

The x-intercept is indeed 7.

The y-intercept occurs when $x = 0$.

$g(x) = \sqrt[3]{0 - 8} + 1$	Substitute 0 for x in the function equation.
$g(x) = \sqrt[3]{-8} + 1$	Simplify within the radical.
$g(x) = -2 + 1$	Evaluate $\sqrt[3]{-8}$.
$g(x) = -1$	Add −2 and 1.

The y-intercept is −1, as we expected.

🔓 1

The function $g(x)$, like the other two cube root functions we have seen so far, is always increasing.

A cube root function of the form $f(x) = a\sqrt[3]{x - b} + c$ is either always increasing or always decreasing. It has a domain of all real numbers and a range of all real numbers. It has exactly one x-intercept and exactly one y-intercept, although in some cases those occur at the same point (the origin).

EXAMPLE 2

Graph the functions $s(x) = 4\sqrt[3]{x}$ and $t(x) = \sqrt[3]{4x}$, and compare them to each other and to $y = \sqrt[3]{x}$, shown previously.

Let's create a table of values for each of these functions. Choose x-values that will be easy to work with when evaluating each of the cube root expressions.

x	$s(x) = 4\sqrt[3]{x}$	$s(x)$
-8	$s(-8) = 4\sqrt[3]{-8} = 4(-2) = -8$	-8
-1	$s(-1) = 4\sqrt[3]{-1} = 4(-1) = -4$	-4
0	$s(0) = 4\sqrt[3]{0} = 4(0) = 0$	0
1	$s(1) = 4\sqrt[3]{1} = 4(1) = 4$	4
8	$s(8) = 4\sqrt[3]{8} = 4(2) = 8$	8

x	$t(x) = \sqrt[3]{4x}$	$t(x)$
-2	$t(-2) = \sqrt[3]{4(-2)} = \sqrt[3]{-8} = -2$	-2
-1/4	$t(-1/4) = \sqrt[3]{4\left(-\dfrac{1}{4}\right)} = \sqrt[3]{-1} = -1$	-1
0	$t(0) = \sqrt[3]{4(0)} = \sqrt[3]{0} = 0$	0
1/4	$t(1/4) = \sqrt[3]{4\left(\dfrac{1}{4}\right)} = \sqrt[3]{1} = 1$	1
2	$t(2) = \sqrt[3]{4(2)} = \sqrt[3]{8} = 2$	2
16	$t(16) = \sqrt[3]{4(16)} = \sqrt[3]{64} = 4$	4

The graphs of $s(x) = 4\sqrt[3]{x}$ (passing through $(-8, -8)$, $(-1, -4)$, $(0, 0)$, $(1, 4)$, and $(8, 8)$) and $t(x) = \sqrt[3]{4x}$ (passing through $(-2, -2)$, $(-1/4, -1)$, $(0, 0)$, $(1/4, 1)$, and $(2, 2)$) are shown below.

Although we are not showing the point $(16, 4)$ on the graph of $t(x)$, it gives us a sense of the rate at which the function is increasing. We could also estimate the value of $t(x)$ at $x = 8$, to approximate the extension of the graph to here: $\sqrt[3]{4(8)} = \sqrt[3]{32}$ ≈ 3.2. The function $t(x)$ passes through a point near $(8, 3.2)$.

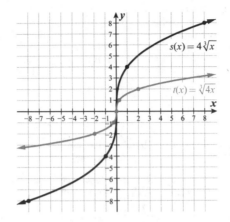

Both of these functions increase more quickly than $y = \sqrt[3]{x}$, so their graphs are more vertically spread out. The graph of $s(x) = 4\sqrt[3]{x}$ increases more quickly than that of $t(x) = \sqrt[3]{4x}$. This makes sense, considering that $\sqrt[3]{4x}$ is the same as $\sqrt[3]{4} \cdot \sqrt[3]{x}$, or approximately $1.59\sqrt[3]{x}$. The value of $4\sqrt[3]{x}$ is greater than $1.59\sqrt[3]{x}$ for all x-values greater than 0.

> The coefficient a of a cube root function $f(x) = a\sqrt[3]{x}$ affects the scale of the graph. For $|a| > 1$, the function increases (or decreases) more quickly than $y = \sqrt[3]{x}$, and for $0 < |a| < 1$, the function increases (or decreases) more slowly. This is also the case for translated versions of this cube root function. When $a < 0$ (when a is negative), the function is reflected across the x-axis.

The graph of $f(x) = a\sqrt[3]{x}$ produces the same image when reflected across the y-axis as when reflected across the x-axis because it is symmetrical with respect to the origin (as is a cubic function of the form $f(x) = ax^3$, as well as any other odd function). If the image is rotated 180°, it exactly matches the original image.

Note that any cube root function of the form $f(x) = a\sqrt[3]{x}$ is an odd function, meaning that $f(-x) = -f(x)$ (or $a\sqrt[3]{-x} = -a\sqrt[3]{x}$). In other words, a reflection of $f(x) = a\sqrt[3]{x}$ across the y-axis produces the same graph as a reflection across the x-axis. However, when the function is of the form $f(x) = a\sqrt[3]{x - b} + c$ and $a < 0$, the translated function is reflected across the line $y = c$ or across the line $x = b$. You can also accomplish this by reflecting the graph across the appropriate axis before translating as indicated, with the same result.

EXAMPLE 3

Graph the function $f(x) = -2\sqrt[3]{x + 5}$.

This function should be the graph of $y = \sqrt[3]{x}$ translated 5 units to the left, reflected across the x-axis (or reflected across the line $x = -5$), and stretched vertically by a factor of 2. Let's create a table of values to find points on the graph.

x	$f(x) = -2\sqrt[3]{x + 5}$	$f(x)$
-13	$f(-13) = -2\sqrt[3]{-13 + 5} = -2\sqrt[3]{-8} = -2(-2) = 4$	4
-6	$f(-6) = -2\sqrt[3]{-6 + 5} = -2\sqrt[3]{-1} = -2(-1) = 2$	2
-5	$f(-5) = -2\sqrt[3]{-5 + 5} = -2\sqrt[3]{0} = -2(0) = 0$	0
-4	$f(-4) = -2\sqrt[3]{-4 + 5} = -2\sqrt[3]{1} = -2(1) = -2$	-2
3	$f(3) = -2\sqrt[3]{3 + 5} = -2\sqrt[3]{8} = -2(2) = -4$	-4

The function $f(x) = -2\sqrt[3]{x + 5}$ passes through the points (-13, 4), (-6, 2), (-5, 0), (-4, -2), and (3, -4), as shown below.

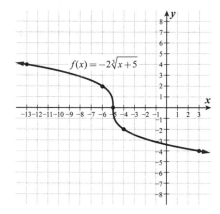

The graph matches our prediction: It is the graph of $y = \sqrt[3]{x}$ translated 5 units to the left, reflected across the x-axis (or across the line $x = -5$), and stretched vertically by a factor of 2.

Notice that this graph is symmetrical with respect to the point $(-5, 0)$, which is the point of inflection for this cube root function graph.

> The point of inflection of a cube root function graph of the form $f(x) = a\sqrt[3]{x - b} + c$ is the point (b, c). The graph is symmetrical with respect to this point.

This is why a reflection across the line $y = c$ produces the same result as a reflection across the line $x = b$.

In Example 3, $b = -5$ and $c = 0$ in the form $f(x) = a\sqrt[3]{x - b} + c$, so a reflection across the line $x = -5$ has the same effect as a reflection across the line $y = 0$, also known as the x-axis.

EXAMPLE 4

If $p(x) = x^3 + 3x^2 - 4x - 12$, graph its inverse relation. Is this inverse relation a cube root function?

For $p(x)$, it is not easy to solve for its inverse. If we set $x = y^3 + 3y^2 - 4y - 12$, we cannot simply solve for the new y. Instead, let's graph $p(x) = x^3 + 3x^2 - 4x - 12$ and then reflect that graph across the line $y = x$.

If we factor x^2 out of the first two terms and -4 out of the last two terms, each pair of remaining terms is $(x + 3)$.

$x^2(x + 3) - 4(x + 3)$	Factor each pair of terms in $x^3 + 3x^2 - 4x - 12$.
$(x^2 - 4)(x + 3)$	Use the distributive property to factor out $(x + 3)$.
$(x - 2)(x + 2)(x + 3)$	Factor $x^2 - 4$, a difference of squares.

So, $p(x) = x^3 + 3x^2 - 4x - 12 = (x - 2)(x + 2)(x + 3)$, so it has x-intercepts of 2, -2, and -3. It is a cubic with a positive leading coefficient, so its left arm points down and its right arm points up. When $x = 0$, $p(x) = -12$, so its y-intercept is -12. We can also use the equation to solve for a few more points on the graph.

$p(-4) = (-4)^3 + 3(-4)^2 - 4(-4) - 12 = -64 + 48 + 16 - 12 = -12$
$p(-1) = (-1)^3 + 3(-1)^2 - 4(-1) - 12 = -1 + 3 + 4 - 12 = -6$
$p(1) = 1^3 + 3(1^2) - 4(1) - 12 = 1 + 3 - 4 - 12 = -12$
$p(3) = 3^3 + 3(3^2) - 4(3) - 12 = 27 + 27 - 12 - 12 = 30$

The graph of $p(x) = x^3 + 3x^2 - 4x - 12$ is shown below.

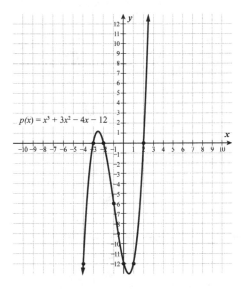

When we reflect this graph across the line $y = x$, each point (a, b) will be mapped to a point (b, a). So, the inverse relation of $p(x)$ will pass through the points (−12, −4), (0, −3), (0, −2), (−6, −1), (−12, 0), (−12, 1), and (0, 2), as shown below.

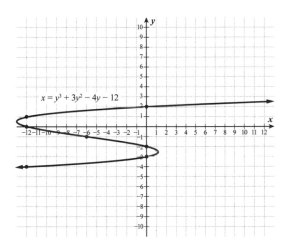

This graph fails the vertical line test, so the relation $x = y^3 + 3y^2 - 4y - 12$ is not a function.

However, if we set constraints on the domain of the function $p(x)$, then its inverse will be a function. The portion of $p(x)$ that is between $x = -2$ and $x = 0$ is always decreasing, meaning that only one x-value is associated with each y-value here, so the section of the graph of the inverse relation between $y = -2$ and $y = 0$ is a function, with only one y-value associated with each x-value. So, if $p(x) = x^3 + 3x^2 - 4x - 12$ for $-2 \leq x \leq 0$, then $p^{-1}(x)$ is a function, as shown below.

Alternatively, we could have defined $p(x)$ to have a domain of only $x \geq 1$, where the graph is always increasing, and this would create an inverse that is also a function. Or, we could have chosen the section where $x \leq -3$.

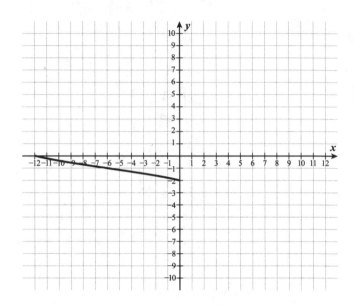

Lesson 6.2
Piecewise-Defined Functions

So far in this book, we have only looked at functions that can be defined by a single polynomial, rational, radical, exponential, logarithmic, or trigonometric function. Some of these functions, such as most rational functions, have discontinuities, with different sections behaving differently, but all have been defined by single functions.

A **piecewise-defined function** is defined by different equations for its various pieces. Pay attention to whether or not endpoints are included in a given section.

EXAMPLE 5

Graph the function defined below. Describe its domain and range.

$$f(x) = \begin{cases} -2x, & x < 0 \\ 3, & 0 \leq x \leq 4 \\ x - 1, & x > 4 \end{cases}$$

The function is $f(x) = -2x$ for x-values less than 0. The x-value 0 is not included in this section, so we must use an open circle at the point (0, 0) to indicate that this endpoint is not included in the function.

The notation "$-2x$, $x < 0$" means "$-2x$ for $x < 0$." The bracket indicates that $-2x$, 3, and $x - 1$ are all definitions of the function $f(x)$, each for a certain portion of the domain, as indicated. In other words, $f(x)$ is defined to be equal to $-2x$ for $x < 0$, to be equal to 3 for $0 \leq x \leq 4$, and to be equal to $x - 1$ for $x > 4$.

The *x*-value of 0 is not included because the inequality symbol is <, not ≤. If 0 were included in the set, then *f*(0) would equal –2(0), which is 0. So, the open circle must go at the point (0, 0) to indicate that this is the endpoint but is not included.

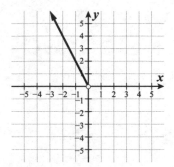

For *x*-values between 0 and 4, *f*(*x*) = 3. This is a horizontal line segment. Both endpoints are included in this section, so use closed circles for the points (0, 3) and (4, 3).

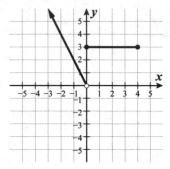

For *x*-values greater than 4, the function is defined as *f*(*x*) = *x* – 1. The *x*-value of 4 is not included, so the point (4, 3) is not included in this section. However, the point (4, 3) is included in the previous section, so that closed circle point remains in our graph. Here is the complete graph of *f*(*x*).

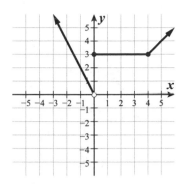

Although this graph has a discontinuity (at $x = 0$), the function is defined for all real numbers. The domain of the function is all real numbers. All function values are above the x-axis, and both ends are continuing forever upward, with the left ray encompassing all positive numbers. So, the range of the function is all real numbers greater than 0.

Piecewise-defined functions may be defined using any kinds of functions, not just linear.

EXAMPLE 6

Graph the function defined below. Describe its domain and range.

$$h(x) = \begin{cases} \dfrac{x}{x-2}, & x \leq 2 \\ -\log_2 x, & x > 2 \end{cases}$$

The function $y = \dfrac{x}{x-2}$ has two unconnected sections, on either side of the vertical asymptote $x = 2$. We are only graphing the left-side section, because $h(x) = \dfrac{x}{x-2}$ only for x-values less than or equal to 2. This curve is always decreasing, with its left end approaching a horizontal asymptote of $y = 1$ and its right end dropping along the vertical asymptote $x = 2$. It passes through (0, 0) and (1, −1).

See Lesson 3.3 for a review of graphing rational functions. In this case, we are only graphing one of the two curves that represent this rational function, because of the domain restrictions given in the function definition.

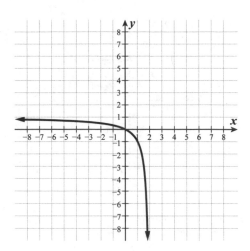

The graph of $y = -\log_2 x$ is a decreasing function, with its left end approaching a vertical asymptote of $x = 0$ and its right end heading toward $-\infty$. However, we only want to graph the portion of it to the right of $x = 2$. This x-value is not included, so we must put an open circle at $(2, -1)$. The graph passes through $(4, -2)$ and $(8, -3)$.

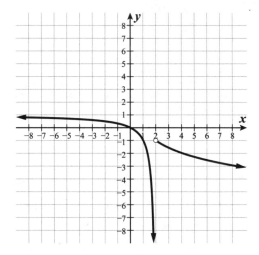

Use the equation $y = -\log_2 x$ to find points along its graph. When $x = 2$, $y = -\log_2 2$ $= -(1) = -1$. When $x = 4$, $y = -\log_2 4 = -(2)$ $= -2$. When $x = 8$, $y = -\log_2 8 = -(3) = -3$.

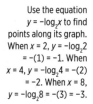

In Example 19 in Lesson 5.4, we graphed $y = \log_2 x$. The graph of $y = -\log_2 x$ is a reflection of that graph across the x-axis, as seen in the portion of $y = -\log_2 x$ shown here.

Even though the function definition encompasses $x \le 2$ and $x > 2$, or all x-values, the function is undefined at $x = 2$. When $x = 2$, $h(x) = \dfrac{x}{x - 2}$, and the expression $\dfrac{x}{x - 2}$ is undefined when $x = 2$. So, the domain of the function $h(x)$ is all real numbers except 2. The function has two arms extending toward $-\infty$, and the other arm is heading toward, but never reaches, a function value of 1. So, the range of $h(x)$ is all real numbers less than 1, or $h(x) < 1$.

STEP FUNCTIONS

> A **step function** is a piecewise-defined function that consists only of horizontal line segments and/or rays.

A step function is discontinuous because the line segments, at various y-values, are not connected to one another.

Technically, a constant function (such as the line $y = 5$) is a continuous step function, consisting of just one line across a domain of all real numbers. But, here we will focus only on discontinuous step functions.

EXAMPLE 7

Graph the function $f(x)$, defined below. Identify its x- and y-intercepts, where they exist. What are the domain and range of $f(x)$?

$$f(x) = \begin{cases} -2, & x < -4 \\ 3, & -4 \leq x \leq 0 \\ 1, & 0 < x \leq 2 \\ -3, & x > 2 \end{cases}$$

For x-values less than -4, $f(x) = -2$, so draw a horizontal ray at $y = -2$ from $(-4, -2)$ to the left. The endpoint is not included ($x < -4$), so use an open circle at the point $(-4, -2)$.

For x-values between -4 and 0, inclusive, $f(x) = 3$, so draw a horizontal line segment at $y = 3$ from $(-4, 3)$ to $(0, 3)$. Both endpoints are included ($x \geq -4$ and $x \leq 0$), so use closed circles at the points $(-4, 3)$ and $(0, 3)$.

For x-values between 0 and 2, $f(x) = 1$, so draw a horizontal line segment at $y = 1$ from $(0, 1)$ to $(2, 1)$. The left endpoint is not included ($x > 0$), but the right endpoint is ($x \leq 2$). Use an open circle at $(0, 1)$ and a closed circle at $(2, 1)$.

For x-values greater than 2, $f(x) = -3$, so draw a horizontal ray at $y = -3$ from $(2, -3)$ to the right. The endpoint is not included ($x > 2$), so use an open circle at $(2, -3)$.

Here is a graph of step function $f(x)$.

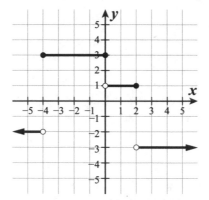

This function never intersects the x-axis, so it has no x-intercept. It intersects the y-axis at 3, so its y-intercept is 3. Notice that it does not actually intersect the y-axis at 1, because there is an open circle at that point.

This step function is continuous along each of its four sections. At the points of discontinuity, the function value is still defined: $f(-4) = 3$, $f(0) = 3$, and $f(2) = 1$. Because the function is defined for all real-number values of x, the domain of $f(x)$ is all real numbers.

The standard way to express a set of values, such as those included in a function range, is in ascending order: {-3, -2, 1, 3}.

The function values, however, only include -2, 3, 1, and -3. The range of $f(x)$ is the set {-3, -2, 1, 3}.

FLOOR AND CEILING FUNCTIONS

In keeping with the architectural theme, let's take a look at floor and ceiling functions.

> The **floor function**, which is written as $f(x) = \lfloor x \rfloor$, is defined as the greatest integer less than or equal to x. It is also known as the greatest integer function.

EXAMPLE 8

Graph the floor function. Identify its x- and y-intercepts, where they exist.

Let's consider various values of x, including those between consecutive integers. When $x = 0$, the greatest integer less than or equal to 0 is 0. When $x = 1/2$, the greatest integer less than or equal to 1/2 is 0. This is true for any fraction greater than 0 and less than 1. When $x = 1$, the greatest integer less than or equal to 1 is 1. Because the definition of the function includes integer values "equal to x," each integer x-value is mapped to itself.

When $x = 1\,1/2$, the greatest integer less than or equal to $1\,1/2$ is 1. This is true for all x greater than 1 and less than 2.

So, the function graph is a series of steps that are each 1 unit long. Each step begins at (x, x) (the point where $f(x) = x$), where x is an integer, and extends horizontally to, but does not include, the point $(x, x + 1)$. This means that the left endpoint of each line segment is a closed circle and the right endpoint of each line segment is an open circle. The graph of $f(x) = \lfloor x \rfloor$ is shown below.

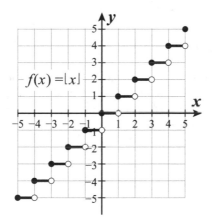

The same pattern extends into the negative *x*-values, as shown in the graph. For example, $\lfloor -1.5 \rfloor = -2$, because the greatest integer less than or equal to −1.5 is −2. The value −1 is greater than −1.5.

The *y*-intercept of the floor function is 0, at the origin. The *x*-intercepts are all the points along the "step" that extends from *x* = 0 to *x* = 1, not including *x* = 1. So, the *x*-intercepts are all *x*-values in the set 0 ≤ *x* < 1.

> The **ceiling function**, which is written as $f(x) = \lceil x \rceil$, is defined as the least integer that is greater than or equal to *x*.

EXAMPLE **9**

The ceiling function $f(x) = \lceil x \rceil$ is graphed below.
What are the domain and range of this function?

Both floor and ceiling functions consist of an infinite number of "steps," because there are an infinite number of integers.

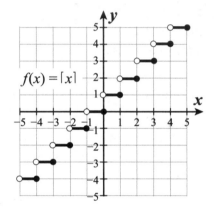

$$f(x) = \lceil x \rceil$$

Even though there is a discontinuity at each integer, the function is defined for every value of *x*. That means that the domain is all real numbers.

The function values include only integers and no other numbers between them. The range is the set of all integers.

Graph the function $p(x) = \lceil x \rceil + 2$.

As with other function types, adding a constant to the function value shifts the function vertically that number of units. In this case, a positive 2 is added to $\lceil x \rceil$, so $p(x)$ is the ceiling function shifted 2 units directly upward.

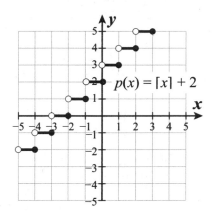

$$p(x) = \lceil x \rceil + 2$$

To check our work, let's evaluate $p(x)$ for a few values of x.

When $x = -1/2$, the least integer that is greater than or equal to $-1/2$ is 0, so $p(-1/2) = 0 + 2 = 2$.

When $x = 1$, the least integer that is greater than or equal to 1 is 1, so $p(1) = 1 + 2 = 3$.

When $x = 2\ 1/4$, the least integer that is greater than or equal to $2\ 1/4$ is 3, so $p(2\ 1/4) = 3 + 2 = 5$.

The points $(-1/2, 2)$, $(1, 3)$, and $(2\ 1/4, 5)$ are all on our graph, so we have correctly graphed $p(x)$.

ABSOLUTE VALUE FUNCTIONS

An absolute value function is a type of piecewise-defined function, because it is defined differently for negative arguments than for positive arguments.

> An **absolute value function** is a function in which the independent variable (x) is within an absolute value expression.

The most basic absolute value function is $f(x) = |x|$. For all positive numbers and 0, the function value is the x-value itself. In other words, for $x \geq 0$, $f(x) = x$. For all negative numbers, the absolute value produces a positive function value, which is the same result we would get by multiplying each negative x-value by −1. In other words, for $x < 0$, $f(x) = -x$.

The graph of $f(x) = |x|$ is the graph of $f(x) = -x$ for $x < 0$ and the graph of $f(x) = x$ for $x \geq 0$, so it has a V shape, as shown below.

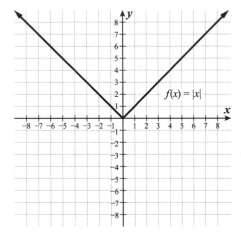

The two arms of the function extend forever along these diagonal lines, each approaching ∞ as x approaches ∞ and −∞. The minimum value of $f(x) = |x|$ is 0. So, the domain is all real numbers, and the range is all real numbers greater than or equal to 0.

These vertical and horizontal translations are similar to what we saw in Lesson 1.4, as well as in Example 17 in Lesson 3.6 and elsewhere throughout this book.

As with other kinds of functions, for absolute value functions, $f(x) + k$ represents a vertical shift of $f(x)$ by k units and $f(x + k)$ represents a horizontal shift of $f(x)$ by k units.

Graph the function $g(x) = |x + 2| - 5$. Then, describe its x- and y-intercepts, domain, range, and line of symmetry.

If $f(x) = |x|$, the function $g(x)$ is $f(x + 2) - 5$, so its graph is the graph of $f(x) = |x|$ shifted 2 units to the left and 5 units down.

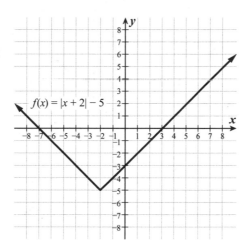

To confirm that this is correctly graphed, let's test values in the given equation and compare the results to points on the graph.

When $x = 0$, $g(x) = |0 + 2| - 5 = |2| - 5 = 2 - 5 = -3$.
When $x = 3$, $g(x) = |3 + 2| - 5 = |5| - 5 = 5 - 5 = 0$.
When $x = -6$, $g(x) = |-6 + 2| - 5 = |-4| - 5 = 4 - 5 = -1$.

The equation produces the coordinate pairs $(0, -3)$, $(3, 0)$, and $(-6, -1)$, all of which are points lying on the graph of $g(x)$, so the graph appears to be correct.

Look at where the graph crosses the x- and y-axes. The function $g(x) = |x + 2| - 5$ has x-intercepts of -7 and 3 and a y-intercept of -3.

The function $g(x)$ is defined for all real numbers, so its domain is all real numbers ($-\infty \leq x \leq \infty$). The minimum value of $g(x)$ is -5, so its range is $g(x) \geq -5$.

Notice that some function values are negative. Even though an absolute value is never negative, an absolute value function may have negative values.

The line of symmetry for this graph is the vertical line $x = -2$, as shown below. The parts of the graph to the left and to the right of this line are mirror reflections of one another.

We shifted the view of the graph over slightly, to better see the symmetry, but it is the same function graph as shown on the previous page.

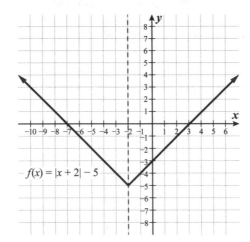

$f(x) = |x + 2| - 5$

The reason that the maximum or minimum occurs when $x = -c/b$ is because this is the value of x that makes $(bx + c)$ equal to 0. The minimum value of an absolute value expression of the form $|bx + c|$ is always 0. So, the minimum or maximum value of $a|bx + c| + d$ is $0 + d$, or d.

An absolute value function of the form $y = a|bx + c| + d$, where a, b, c, and d are constants, is always a V shape (or an upside-down V shape, in the case that a is negative). So, for this kind of absolute value function:

• The turning point of the V, called the **vertex**, is the maximum or minimum value of the function.

• The domain is all real numbers, and the range is all real numbers less than the maximum or greater than the minimum (depending on whether the function has a maximum or minimum).

• There are 0, 1, or 2 x-intercepts and exactly 1 y-intercept.

• The graph is symmetrical with respect to a vertical line at $x = -c/b$. The vertex lies on this line of symmetry, with an x-coordinate of $-c/b$ and a y-coordinate of d. This means that the maximum or minimum value of the function is equal to d.

• Both arms of the graph are approaching ∞ or both arms of the graph are approaching $-∞$.

Another way to graph an absolute value function is by graphing each of its two halves: the ray to the left of its vertex and the ray to the right. Use the function equation to determine the equation of the line along which each ray lies.

For $g(x) = |x + 2| - 5$, write the equations for the case where $(x + 2)$ is positive and the case where $(x + 2)$ is negative. When $(x + 2)$ is positive, $g(x) = (x + 2) - 5$, which simplifies to $g(x) = x - 3$. When $(x + 2)$ is negative, we must multiply that value by -1 to produce its absolute value: $g(x) = -(x + 2) - 5$, which simplifies to $g(x) = -x - 7$.

So, for positive values of $(x + 2)$, which are x-values greater than -2, $g(x) = x - 3$, and for negative values of $(x + 2)$, which are x-values less than -2, $g(x) = -x - 7$. This confirms our graphing work above; the half of the V to the right of $x = -2$ follows the line $y = x - 3$, and the half of the V to the left of $x = -2$ follows the line $y = -x - 7$.

While adding a constant to some part of the equation $f(x) = |x|$ results in a translation, as in Example 11 above, multiplying by a constant results in a vertical/horizontal stretch and/or reflection.

If you'd like to see how absolute value questions appear on the SAT, check out the online Student Tools for this book.

EXAMPLE

Graph the function $f(x) = -3|x - 4|$ and describe its domain and range. What is the line of symmetry for this graph?

The graph of $f(x) = -3|x - 4|$ should be the graph of $y = |x|$ shifted 4 units to the right, reflected across the x-axis, and stretched vertically. Let's create a table of values to find points on the graph.

x	$f(x) = -3\|x - 4\|$	$f(x)$
0	$f(0) = -3\|0 - 4\| = -3\|-4\| = -3(4) = -12$	-12
1	$f(1) = -3\|1 - 4\| = -3\|-3\| = -3(3) = -9$	-9
2	$f(2) = -3\|2 - 4\| = -3\|-2\| = -3(2) = -6$	-6
3	$f(3) = -3\|3 - 4\| = -3\|-1\| = -3(1) = -3$	-3
4	$f(4) = -3\|4 - 4\| = -3\|0\| = -3(0) = 0$	0
5	$f(5) = -3\|5 - 4\| = -3\|1\| = -3(1) = -3$	-3
6	$f(6) = -3\|6 - 4\| = -3\|2\| = -3(2) = -6$	-6

Here is the graph of $f(x) = -3|x - 4|$, passing through the points (0, –12), (1, –9), (2, –6), (3, –3), (4, 0), (5, –3), and (6, –6).

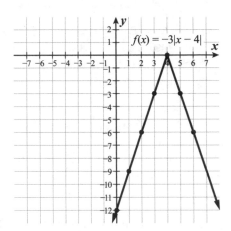

As expected, the graph is $y = |x|$ shifted 4 units to the right, reflected across the x-axis, and stretched vertically by a factor of 3.

For $f(x) = -3|x - 4|$, the domain is $-\infty \leq x \leq \infty$ and the range is $f(x) \leq 0$.

The graph of $f(x)$ has a vertical line of symmetry of $x = 4$. Each point on the function graph is equidistant from this line of symmetry as a point horizontally aligned with it on the other side of the line.

⑫

Let's use the function equation to find the equation for each ray of the graph, as an alternative approach to graphing this absolute value function.

We could also use the formula $(-c/b, d)$ to find that the maximum has the coordinates (4, 0).

First, find the x-value of the maximum, when the argument of the absolute value is 0: in this case, when $x - 4 = 0$. The maximum occurs when $x = 4$, and the maximum is $f(4) = -3|4 - 4| = -3(0) = 0$.

When $x < 4$, $f(x) = -3(-1)(x - 4) = 3(x - 4) = 3x - 12$.
When $x > 4$, $f(x) = -3(x - 4) = -3x + 12$.

As shown in our graph above, the left ray lies on the line $y = 3x - 12$ and the right ray lies on the line $y = -3x + 12$.

Notice that the slopes of the two rays of the absolute value function graph are opposites: 3 and −3 in Example 12 and 1 and −1 in Example 11. Because the V shape of an absolute value function of the form $y = a|bx + c| + d$ is always symmetrical with respect to a vertical line through the vertex, the two rays of the function will always have slopes that are opposites of one another.

EXAMPLE 13

Graph the function $p(x) = |5 - x| + 3$.

This function can also be written as $p(x) = |-(x - 5)| + 3$, so its graph is $y = |x|$ shifted 5 units to the right and 3 units up, then reflected across the line $x = 5$.

Algebraically, for $x \le 5$, $p(x) = (5 - x) + 3 = -x + 8$, and for $x > 5$, $p(x) = -(5 - x) + 3 = x - 2$. So, to the left of the vertex (5, 3), the graph follows the line $y = -x + 8$, and to the right of the vertex, the graph follows the line $y = x - 2$.

The reflection across the line of symmetry is unnecessary, because it produces an identical graph. Algebraically, the absolute value expressions $|5 - x|$ and $|x - 5|$ are equivalent, because they each represent the positive difference between x and 5, or the distance between x and 5 on a number line. So, $p(x) = |5 - x| + 3$ is the same as $p(x) = |x - 5| + 3$, which is the graph of $y = |x|$ translated 5 units to the right and 3 units up.

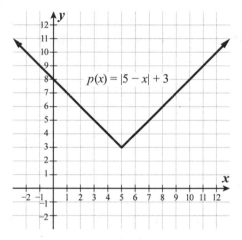

The algebraic process that we use to determine the lines of the graph of an absolute value function is the same process we use to solve an absolute value equation: Replace the absolute value with its own argument for one equation and with the opposite of that expression in the second equation. Solve both equations to determine the solutions to the equation, if any exist.

For example, to solve the equation $|5 - x| + 3 = 7$, we would write the equations $(5 - x) + 3 = 7$ and $-(5 - x) + 3 = 7$.

$$5 - x + 3 = 7$$
$$-x + 8 = 7$$
$$x = 1$$

$$-5 + x + 3 = 7$$
$$x - 2 = 7$$
$$x = 9$$

The two solutions to the equation $|5 - x| + 3 = 7$ are $x = 1$ and $x = 9$. We could also determine these solutions by looking at the graph in Example 13. The points where $p(x) = 7$ occur when $x = 1$ and when $x = 9$.

EXAMPLE 14

Find all solutions to the equation $|x - 2| = x^2 - 2x - 2$.

Let's solve algebraically for the cases where $x - 2 \geq 0$ and where $x - 2 < 0$. In each case, we must make sure that our solutions match these constraints.

When $x \geq 2$, the absolute value equation becomes $x - 2 = x^2 - 2x - 2$.

$x = x^2 - 2x$	Add 2 to both sides.
$0 = x^2 - 3x$	Subtract x from both sides.
$0 = x(x - 3)$	Factor the quadratic.
$x = 0, x = 3$	Solve for each factor set equal to 0.

Absolute value equations sometimes produce extraneous solutions, similar to rational equations, as we saw in Lesson 3.1, and square root equations, as we saw in Lesson 3.4.

However, our constraint for this part of the solution set is that $x \geq 2$. The equation $x - 2 = x^2 - 2x - 2$ is only true when $x \geq 2$. So, $x = 0$ is an extraneous solution. For $x \geq 2$, the solution to the equation is $x = 3$.

When $x < 2$, the absolute value equation becomes $-(x - 2) = x^2 - 2x - 2$.

$-x + 2 = x^2 - 2x - 2$	Distribute the negative sign to both terms in parentheses.
$0 = x^2 - x - 4$	Add x to, and subtract 2 from, both sides.

$$x = \frac{-(-1) \pm \sqrt{(-1)^2 - 4(1)(-4)}}{2(1)}$$

Use the quadratic formula, because we cannot easily factor.

$$x = \frac{1 \pm \sqrt{17}}{2}$$

Simplify.

A solution for this section only applies when $x < 2$, so $x = \dfrac{1 + \sqrt{17}}{2}$ is an extraneous

solution. For $x < 2$, the solution to the equation is $x = \dfrac{1 - \sqrt{17}}{2}$.

The two solutions to the equation $|x - 2| = x^2 - 2x - 2$ are $x = 3$ and $x = \dfrac{1 - \sqrt{17}}{2}$,

which has an approximate value of -1.56.

To quickly tell whether $\dfrac{1 + \sqrt{17}}{2}$ is less than or greater than 2, use an approximation for $\sqrt{17}$. The value of $\sqrt{17}$ is a little greater than $\sqrt{16}$, or 4. Using 4 in place of $\sqrt{17}$ in $\dfrac{1 + \sqrt{17}}{2}$ gives us $\dfrac{1 + 4}{2}$, which simplifies to 5/2, or 2.5. So, the value of $\dfrac{1 + \sqrt{17}}{2}$ is slightly greater than 2.5, which means that it is definitely greater than 2.

If we graph $y = |x - 2|$ and $y = x^2 - 2x - 2$ on the same coordinate grid, we can identify the points of intersection of the two graphs. The x-values of these points represent solutions to the original equation, because they produce the same value for $|x - 2|$ as for $x^2 - 2x - 2$ (the same y-value on each graph).

The equation $y = x^2 - 2x - 2$ in vertex form is $y = (x - 1)^2 - 3$, so this is the parabola $y = x^2$ shifted 1 unit right and 3 units down, with its vertex at $(1, -3)$.

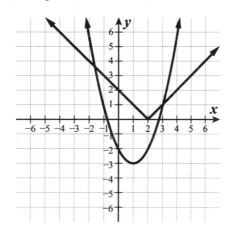

In Lesson 2.2, we looked at how you can solve a polynomial equation using a system of equations. Here we are applying the same concept, but one of the equations in the system is an absolute value function.

The x-values of the points of intersection are 3 and about −1.56, so the solutions to the equation $|x - 2| = x^2 - 2x - 2$ are $x = 3$ and $x = \dfrac{1 - \sqrt{17}}{2}$, as we found above, algebraically.

The extraneous solutions we found are the x-values of the points where the parabola $y = x^2 - 2x - 2$ would intersect with the extensions of each of the rays of $y = |x - 2|$ into full lines. For example, the complete line $y = x - 2$ would intersect the parabola $y = x^2 - 2x - 2$ at the points (0, −2) and (3, 1), but the absolute value function $y = |x - 2|$ does not include the point (0, −2).

Lesson 6.3
Exponential Functions

An **exponential function** is a function in which the independent variable (x) is in the exponent of a constant base that is greater than 0 and not equal to 1. The most basic form of an exponential function is $f(x) = b^x$ (with $b > 0$ and $b \neq 1$).

Exponential functions were briefly introduced in Lesson 5.4, as the inverse of logarithmic functions. Here we will explore exponential functions in greater detail.

The base must be positive, because a negative base would alternately produce positive and negative function values respectively for even and odd values of x. The base cannot equal 0, because the expression 0^x is equal to 0, regardless of the value of x, creating a constant function instead of an exponential function. Likewise, a base of 1 in $f(x) = b^x$ creates the constant function $f(x) = 1$, because 1 raised to any power is just 1.

Graph the exponential functions $g(x) = 3^x$ and $h(x) = (1/3)^x$. What is the relationship between the two graphs?

First, create a table of values for each function. Use x-values that provide easy function calculations.

x	$g(x) = 3^x$	$g(x)$
-2	$g(-2) = 3^{-2} = \dfrac{1}{3^2} = 1/9$	1/9
-1	$g(-1) = 3^{-1} = 1/3$	1/3
0	$g(0) = 3^0 = 1$	1
1	$g(1) = 3^1 = 3$	3
2	$g(2) = 3^2 = 9$	9

x	$h(x) = (1/3)^x$	$h(x)$
-2	$h(-2) = (1/3)^{-2} = 3^2 = 9$	9
-1	$h(-1) = (1/3)^{-1} = 3^1 = 3$	3
0	$h(0) = (1/3)^0 = 1$	1
1	$h(1) = (1/3)^1 = 1/3$	1/3
2	$h(2) = (1/3)^2 = 1/9$	1/9

The graphs of $g(x) = 3^x$ (passing through (-2, 1/9), (-1, 1/3), (0, 1), (1, 3), and (2, 9)) and $h(x) = (1/3)^x$ (passing through (-2, 9), (-1, 3), (0, 1), (1, 1/3), and (2, 1/9)) are shown on the coordinate grid below.

Remember that any value b raised to the power of −1 is $1/b$. So, 3^{-2}, which is the same as $(3^2)^{-1}$, is equal to $\dfrac{1}{3^2}$.

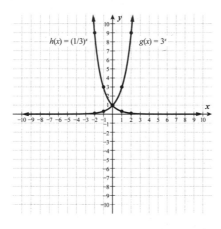

Because $1/3 = 3^{-1}$, the function $h(x) = (1/3)^x$ can be rewritten as $h(x) = 3^{-x}$. This is the function $g(x) = 3^x$ with x replaced by $-x$, or $g(-x)$. The graph of $g(-x) = 3^{-x}$ (also known as $h(x)$) is the reflection of $g(x) = 3^x$ across the y-axis.

The graphs are reflections of one another across the y-axis.

The function $g(x) = 3^x$ is always increasing, and the function $h(x) = (1/3)^x$ is always decreasing.

An exponential function is either always increasing or always decreasing. A function of the form $f(x) = b^x$ is always increasing if $b > 1$ and always decreasing if $0 < b < 1$.

An exponential function in the form $f(x) = b^x$ has a function value of 1 when $x = 0$. In other words, any function of the form $f(x) = b^x$ passes through the point (0, 1).

EXAMPLE 16

Using graphing technology, compare the graphs of $p(x) = 2^x$, $q(x) = 3 \cdot 2^x$, and $r(x) = (3 \cdot 2)^x$.

The function $r(x) = (3 \cdot 2)^x$ can be rewritten as $r(x) = 6^x$.

The graphs of $p(x) = 2^x$, $q(x) = 3 \cdot 2^x$, and $r(x) = 6^x$ are shown below.

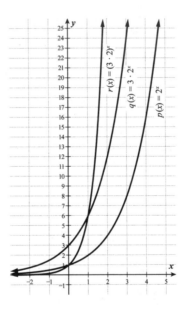

Both $q(x) = 3 \cdot 2^x$ and $r(x) = (3 \cdot 2)^x$ increase in value more quickly than $p(x) = 2^x$, with function graphs that appear more narrow. The function $r(x) = (3 \cdot 2)^x$ increases more quickly than $q(x) = 3 \cdot 2^x$ to the right of $x = 1$. This makes sense, because $(3 \cdot 2)^x$ is the same as $3^x \cdot 2^x$. For x-values greater than 1, 3^x will be greater than 3, so $3^x \cdot 2^x$ will be greater than $3 \cdot 2^x$.

Both $p(x) = 2^x$ and $r(x) = (3 \cdot 2)^x$ (or 6^x) have a y-intercept of 1, as do all functions of the form $f(x) = b^x$, but $q(x) = 3 \cdot 2^x$ has a y-intercept of 3. Its exponential component, 2^x, is equal to 1 when $x = 0$, but that value gets multiplied by 3 in the function.

> An exponential function of the form $f(x) = a \cdot b^x$
> has a *y*-intercept of (0, *a*). For |*a*| > 1, it increases
> (or decreases) more quickly than $f(x) = b^x$, and for
> 0 < |*a*| < 1, it increases (or decreases) more slowly. A negative
> value of *a* reflects the function across the *x*-axis.

All of the exponential functions we have looked at so far have had one arm approaching a horizontal asymptote of *y* = 0, or the *x*-axis. However, a vertical shift of any of these graphs would result in a different horizontal asymptote.

An exponential function of the form $y = a \cdot b^x$ always has a horizontal asymptote of the *x*-axis, because b^x approaches 0 for one end of the graph (as *x* approaches −∞, if *b* > 1; and as *x* approaches ∞, if *b* < 1).

EXAMPLE

Graph the function $f(x) = 1/2 \cdot 2^x - 8$. What are the domain, range, *x*-intercept, and *y*-intercept of this function?

This function should increase at half the level of $y = 2^x$, but also shift vertically downward 8 units. Let's create a table of values to find points on the graph.

x	$f(x) = 1/2 \cdot 2^x - 8$	$f(x)$
0	$f(0) = 1/2 \cdot 2^0 - 8 = 1/2 \cdot 1 - 8 = 1/2 - 8 = -7\,1/2$	−7 1/2
1	$f(1) = 1/2 \cdot 2^1 - 8 = 1/2 \cdot 2 - 8 = 1 - 8 = -7$	−7
2	$f(2) = 1/2 \cdot 2^2 - 8 = 1/2 \cdot 4 - 8 = 2 - 8 = -6$	−6
3	$f(3) = 1/2 \cdot 2^3 - 8 = 1/2 \cdot 8 - 8 = 4 - 8 = -4$	−4
4	$f(4) = 1/2 \cdot 2^4 - 8 = 1/2 \cdot 16 - 8 = 8 - 8 = 0$	0
5	$f(5) = 1/2 \cdot 2^5 - 8 = 1/2 \cdot 32 - 8 = 16 - 8 = 8$	8

The horizontal asymptote for $y = 1/2 \cdot 2^x$ would be at *y* = 0. The graph of $f(x) = 1/2 \cdot 2^x - 8$ is the graph of $y = 1/2 \cdot 2^x$ shifted 8 units down, so $f(x)$ has a horizontal asymptote of *y* = −8.

The graph of $f(x) = 1/2 \cdot 2^x - 8$, with a horizontal asymptote of $y = -8$ and passing through the points $(0, -7\ 1/2)$, $(1, -7)$, $(2, -6)$, $(3, -4)$, $(4, 0)$, and $(5, 8)$, is shown below.

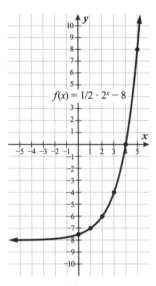

The domain of $f(x) = 1/2 \cdot 2^x - 8$ is all real numbers, and its range is all real numbers greater than -8.

We found both the y-intercept and the x-intercept when creating our table of values. The point $(0, -7\ 1/2)$ is the y-intercept of $f(x)$, and the point $(4, 0)$ is the x-intercept.

Any exponential function of the form $f(x) = a \cdot b^x + c$ has a domain of all real numbers and a range of all real numbers greater than c (in the case that a is positive) or all real numbers less than c (in the case that a is negative). One arm extends toward either positive or negative infinity, and the other arm approaches a horizontal asymptote of $y = c$.

As with other types of functions, $f(x + k)$ represents a horizontal shift of $f(x)$ a total of k units: to the left if k is positive and to the right if k is negative. However, because of the nature of exponential functions, the function equation can be rewritten to eliminate the constant added to x in the exponent, by adjusting the value of a.

EXAMPLE

What equation describes the exponential function, base 3, shown below?

The expression b^{x+k} is equal to $b^x \cdot b^k$, so if b and k are both constants, b^k is also a constant and can be moved out front as part of the coefficient of b^x.

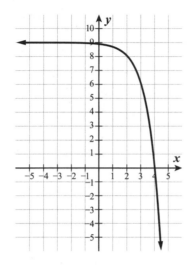

The function has a horizontal asymptote at $y = 9$, so c in $y = a \cdot b^x + c$ is equal to 9. We were also given that $b = 3$. So, the equation is in the form $y = a \cdot 3^x + 9$.

The function is always decreasing, so a is negative.

Let's use points along the graph to determine the exact value of a. The graph passes through the points (2, 8), (3, 6), and (4, 0), so we can substitute each pair of coordinates into the equation $y = a \cdot 3^x + 9$.

$$8 = a \cdot 3^2 + 9 \qquad\qquad 6 = a \cdot 3^3 + 9 \qquad\qquad 0 = a \cdot 3^4 + 9$$
$$8 = 9a + 9 \qquad\qquad 6 = 27a + 9 \qquad\qquad 0 = 81a + 9$$
$$-1 = 9a \qquad\qquad -3 = 27a \qquad\qquad -9 = 81a$$
$$-1/9 = a \qquad\qquad -1/9 = a \qquad\qquad -1/9 = a$$

We could have used just one coordinate pair to solve for a, but using two or three helps confirm that we have solved correctly.

The value of a is $-1/9$, so the equation of this exponential function is $y = -1/9 \cdot 3^x + 9$.

Alternatively, once we determined that the equation is in the form $y = a \cdot 3^x + 9$ with $a < 0$, we could compare the given graph to the graph of $y = -3^x + 9$, which is shown below.

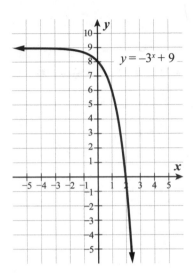

The graph of $y = -3^x + 9$ passes through $(0, 8)$, $(1, 6)$, and $(2, 0)$, and it has a horizontal asymptote of 9. The given graph is therefore the graph of $y = -3^x + 9$ shifted 2 units to the right. So, the equation of the mystery function must be $y = -3^{x-2} + 9$. This equation looks different from what we found before, but they are actually the same equation, as shown below.

$y = -(3^{x-2}) + 9$ Add parentheses to isolate the exponential term.

$y = -(3^x \cdot 3^{-2}) + 9$ Rewrite 3^{x-2} using the rule that $b^{x+y} = b^x \cdot b^y$.
$y = -(3^x \cdot 1/9) + 9$ Evaluate 3^{-2}.
$y = -1/9 \cdot 3^x + 9$ Use the commutative property of multiplication to switch 3^x and $1/9$.

This is the same equation we found above.

EXPONENTIAL FUNCTIONS IN THE REAL WORLD

Exponential functions in the real world are modeling either exponential growth or exponential decay. Most often, an exponential function representing a real-life situation will use time values as the input values (the x-values). The function values typically represent a quantity of some sort (such as mass, population, area, or money), so they are usually positive values, which means that a function of the form $f(x) = a \cdot b^x$ representing the situation has a positive value of a.

In most real-life situations, there is no constant added to $a \cdot b^x$. Decay functions are heading toward an asymptote of 0. Growth functions are shown in relation to some set initial amount or population, but that is given by the variable a in $f(x) = a \cdot b^x$. When $x = 0$, $b^x = 1$, so the function value is equal to a at time = 0.

A function of the form $f(x) = a \cdot b^x$, with $a > 0$, represents exponential growth if $b > 1$ and exponential decay if $0 < b < 1$.

Alternatively, exponential decay can be expressed as $f(x) = a \cdot b^{-x}$, with $a > 0$ and $b > 1$. Remember, $b^{-x} = (1/b)^x$, so if $b > 1$, both of these expressions represent exponential decay.

Exponential functions representing continuous growth or decay often use e for the base.

In Lesson 5.5, Example 24, we looked at an example of exponential decay (of the weight of iodine-131 in relation to time), and the exponential function representing that decay had a base of 1/2 raised to the power of x.

Here is how you may see exponential functions on the SAT.

The population of a certain songbird in North America is decreasing at a rate of 5% per year. If there are currently 450,000 of these birds in North America, and x represents the number of years, which of the following expressions best represents the trend in the North American population of this songbird?

A) $450,000(0.05)x$

B) $450,000(0.05)^x$

C) $450,000(0.95)x$

D) $450,000(0.95)^x$

Jeremy has $2,000 to invest. He is trying to choose between an account at Savers Bank, which offers 1.5% interest compounded continuously, and Cash Cow Bank, which offers 1.0% interest compounded biannually for balances under $2,500 and 2.0% interest compounded biannually for balances of $2,500 or more. (The interest rate changes automatically when an existing balance reaches $2,500.)

Investing a principal, P, in an account with an interest rate of r (expressed as a decimal) compounded n times per year for t years results in a total balance, A, given by the formula $A = P(1 + r/n)^{nt}$. If the interest is compounded continuously, the formula is $A = Pe^{rt}$.

In Lesson 6.2, we learned about piecewise-defined functions. Jeremy's account balance in the Cash Cow Bank account is an example of a real-life situation involving a piecewise-defined function.

Write and graph a function representing the balance over the next 50 years in a Savers Bank account if Jeremy were to invest his $2,000 there. Write a piecewise-defined function representing the balance in a Cash Cow Bank account if Jeremy were to invest his $2,000 there. (Assume in both cases that he will make no further deposits or withdrawals.) Graph the piecewise-defined function on the same coordinate grid as the Savers Bank account function. For what period of investment time is the Savers Bank account a better deal for Jeremy? For what period of investment time is the Cash Cow Bank account a better deal? (Answer to the nearest whole year.)

Let's start by using the continuously compounded interest formula, $A = Pe^{rt}$, to write an equation representing Jeremy's potential investment at Savers Bank. For an investment of $2,000 at an interest rate of 1.5% compounded continuously, $P = 2000$ and $r = 0.015$, so the function is $A = 2000e^{0.015t}$. Using graphing technology, we can graph this exponential function for a period of 50 years, as shown on the coordinate grid below.

Convert 1.5% to a decimal by dividing by 100, or moving the decimal point two places to the left: 1.5% = 0.015.

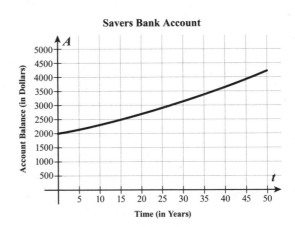

Savers Bank Account

Account Balance (in Dollars) vs. *Time (in Years)*

Jeremy's potential investment in a Cash Cow Bank account is modeled by a piecewise-defined function because the interest rate, and therefore the function equation, changes when the balance reaches $2,500. This is an A-value, so we will need to solve for what the t-value is when A reaches 2500.

First, though, we must write an equation for the account balance when the interest rate is still 1.0% ($r = 0.01$). Again, the initial investment, P, is equal to 2000. The interest for the Cash Cow Bank account is compounded biannually, which means that it is compounded twice per year, so $n = 2$. Using the formula $A = P(1 + r/n)^{nt}$, we get $A = 2000(1 + 0.01/2)^{2t}$, which simplifies to $A = 2000(1.005)^{2t}$.

Now we must find the time, t, when the interest rate changes. This happens when the account balance reaches $2,500, so substitute 2500 for A and solve for t.

$2500 = 2000(1.005)^{2t}$	Substitute 2500 for A in $A = 2000(1.005)^{2t}$.	In Lesson 5.3, we learned how to use the logarithmic change of base property. Here we make use of it to perform calculations related to a real-life situation involving interest rates.
$1.25 = (1.005)^{2t}$	Divide both sides by 2000.	
$\log_{1.005} 1.25 = 2t$	Take the logarithm, base 1.005, of both sides.	
$\dfrac{\log 1.25}{\log 1.005} = 2t$	Use the change of base property to rewrite with common logarithms.	
$2t \approx 44.74$	Use your calculator to evaluate $\dfrac{\log 1.25}{\log 1.005}$.	
$t \approx 22.37$	Divide both sides by 2.	

So, the account balance will reach $2,500 after about 22.37 years, or about 22 years and 4.4 months. However, the interest is only compounded twice per year, and the total number of times it is compounded must be a whole number. In the function $A = 2000(1.005)^{2t}$, the expression $2t$ represents the number of times interest is compounded, so $2t$ must be a whole number. Our solution for $2t$ above, 44.74, must be rounded up to the next whole number, 45, so our solution becomes $t = 22.5$. After the compounding that occurs after 22 1/2 years (the 45th compounding), the account balance will be over $2,500.

To find 5% of the current number of these songbirds, you would multiply 450,000 by 0.05. However, that tells you how many fewer songbirds there will be one year from now. Instead, we want to write an equation that gives the resulting population after x years. If the number decreases by 5% each year, then each year's population is 95% of the previous year's population.

For each increase of x by 1 (for each year that passes), the bird population is multiplied by 0.95. (To find 95% of a value, multiply by 0.95.) A repeated multiplication by the same number can be expressed using an exponential term. The number is the base, and the exponent tells you how many times it is used as a factor. So, multiplying the initial songbird population by 0.95 a total of x times is written as $450,000(0.95)^x$. An exponential function for $f(x)$, the songbird population after x years, is $f(x) = 450,000(0.95)^x$. The correct answer is (D).

For $t = 22.5$, the equation $A = 2000(1.005)^{2t}$ becomes $A = 2000(1.005)^{45}$. Use your calculator to compute this amount. The balance, A, in the account at $t = 22.5$ is $2,503.24, rounded to the nearest cent.

We cannot round the value of $2t$ down to 44, because the account balance will not quite have reached $2,500 at that point.

At this point, the account rules change to a 2.0% interest rate compounded biannually. For the new initial balance (for this equation) of $2,503.24, the formula $A = P(1 + r/n)^{nt}$ produces the equation $A = 2503.24(1.01)^{2t}$. However, this "restarts" the time at 0, whereas really 22.5 years will have passed. We need to shift this function graph 22.5 units (years) to the right, which we can do by subtracting 22.5 from t within the equation. The equation becomes $A = 2503.24(1.01)^{2(t - 22.5)}$, which simplifies to $A = 2503.24(1.01)^{2t - 45}$. So, the account balance at Cash Cow Bank can be modeled by the function below.

$$A = \begin{cases} 2000\,(1.005)^{2t}, & t \leq 22.5 \\ 2503.24\,(1.01)^{2t-45}, & t > 22.5 \end{cases}$$

Let's graph this piecewise-defined function on the same grid as the Savers Bank account graph, using graphing technology.

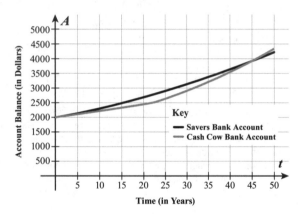

The two function graphs begin at the same point (0, 2000) and appear to intersect when t is approximately equal to 45.5. From $t = 0$ to $t = 45.5$, the balance in Savers Bank is greater, and from $t = 45.5$ to the right, the balance in Cash Cow Bank is greater. So, if Jeremy is going to leave his investment in the bank for any time period up to 45 years, the Savers Bank account will earn him more interest. If he is going to leave his investment in the bank for 46 or more years, the Cash Cow Bank account will earn him more interest.

CHAPTER 6 PRACTICE QUESTIONS

Directions: Complete the following open-ended problems as specified by each question stem. For extra practice after answering each question, try using an alternative method to solve the problem or check your work.

1. Describe the domain, range, end behavior, x-intercept, y-intercept, and point of inflection for the graph of $f(x) = -\sqrt[3]{x - 6} + 2$.

2. Graph the function $h(x) = \lfloor 2x \rfloor$ for $-1 \le x \le 2$. What is the range for this function?

3. If $g(x) = x^3 - 5x^2 + 3x + 9$, what are the x- and y-intercepts of its inverse relation? What is the minimum value of a for which the inverse relation of $g(x)$ with $x \ge a$ is a cube root function?

4. Graph the function $f(x) = -4 \cdot 2^x + 16$.

5. An absolute value function, $r(x)$, has a maximum at $(-4, 3)$ and a left arm with slope 3/2. What equation describes $r(x)$?

6. Find all solutions to the equation $1/2 \, x^2 - 5 = |x - 1|$.

7. At Perry's Pretzel Parlor, employees are paid a starting hourly wage of $12 for their first year working there. After that, their hourly wage increases by 10% each year, on the anniversary of their date of hire. The employee who has worked at Perry's Pretzel Parlor the longest has been there for 4 1/2 years. Write a function that gives $p(x)$, an employee's hourly wage, x years after bring hired at Perry's Pretzel Parlor, for x-values up to 4.5. What kind of function is this?

8. Naima was riding her skateboard at a constant speed of 0.2 mile per minute along a flat stretch of road. After riding 1.2 miles, she hit a downhill stretch, at which point her distance, d, in miles traveled downhill was given by $d = 0.1m^2 + 0.4m$, where m was the number of minutes since she started going downhill. Write and graph a piecewise-defined function that describes $s(x)$, her total distance traveled, throughout her entire 8-minute ride, where x is the number of minutes since she started the ride.

9. The equation $N = A \cdot e^{kt}$ represents the continuous exponential decay of a radioisotope, where N represents the remaining quantity after t years, of an original quantity A, and k is a constant specific to a particular radioisotope. If 77% of an original sample of the radioisotope silicon-32 remains after 65 years, what is the half-life of silicon-32? (The half-life is the amount of time it takes half of the original quantity of a substance to decay.)

SOLUTIONS TO CHAPTER 6 PRACTICE QUESTIONS

1. **domain: all real numbers; range: all real numbers; end behavior: always decreasing; x-intercept: 14; y-intercept: 3.8; point of inflection: (6, 2)**

   A cube root function always has a domain of all real numbers and a range of all real numbers, so this is the case for $f(x)$. A function in this form is either always increasing or always decreasing. In this case, the coefficient of the cube root is negative (-1), so $f(x)$ is always decreasing. This means that its left arm approaches ∞ and its right arm approaches $-\infty$. Its x-intercept occurs when $f(x) = 0$.

$0 = -\sqrt[3]{x - 6} + 2$	Substitute 0 for $f(x)$.
$\sqrt[3]{x - 6} = 2$	Add $\sqrt[3]{x - 6}$ to both sides.
$x - 6 = 8$	Cube both sides.
$x = 14$	Add 6 to both sides.

   The function $f(x) = -\sqrt[3]{x - 6} + 2$ has an x-intercept of 14.

   The y-intercept occurs when $x = 0$.

$f(0) = -\sqrt[3]{0 - 6} + 2$	Substitute 0 for x.
$= -\sqrt[3]{-6} + 2$	Subtract within the radicand.
$\approx -(-1.8) + 2$	Use a calculator to find $\sqrt[3]{-6}$ and round off.
$\approx 1.8 + 2$	A negative multiplied by a negative is a positive.
≈ 3.8	Add 1.8 + 2.

   The function $f(x) = -\sqrt[3]{x - 6} + 2$ has a y-intercept of about 3.8.

   The point of inflection of a function of the form $f(x) = a\sqrt[3]{x - b} + c$ is (b, c), so the point of inflection for $f(x) = -\sqrt[3]{x - 6} + 2$ is $(6, 2)$.

2. **see graph on following page; range: {−2, −1, 0, 1, 2, 3, 4}**

   The symbol ⌊ ⌋ indicates that this is a floor function, or the greatest integer less than or equal to the value of its argument ($2x$ in this case). When $x = -1$, $2x = -2$, and $\lfloor -2 \rfloor = -2$. When $x = 0$, $2x = 0$, and $\lfloor 0 \rfloor = 0$. This is a jump of 2 units, but a floor function for a continuous range of several units should have values at each consecutive integer. The fractional x-value of $-1/2$ produces an $h(x)$-value of -1. All x-values greater than -1 and less than $-1/2$ produce values of $2x$ that are greater than -2 and less than -1, so they produce $h(x)$-values of -2.

   All x-values greater than or equal to $-1/2$ and less than 0 produce an $h(x)$-value of -1. When $x = 0$, $h(x) = 0$. When $x = 1/2$, $h(x) = 1$. For x-values greater than 0 and less than $1/2$, $h(x) = 0$. This pattern continues, with steps each $1/2$ unit long at each integer height, with a closed circle as the left endpoint and an open circle as the right endpoint. The greatest x-value in the domain of $h(x)$ is 2, and $h(2) = \lfloor 2(2) \rfloor = 4$. The graph of $h(x) = \lfloor 2x \rfloor$ for $-1 \le x \le 2$ is shown below.

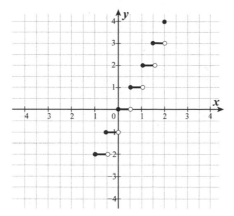

The domain of $h(x)$ was given as $-1 \leq x \leq 2$. The range is each integer from -2 to 4 and does not include any values in between. In other words, the range of $h(x)$ is $\{-2, -1, 0, 1, 2, 3, 4\}$.

3.   **3**

Each coordinate pair (a, b) in a function translates to the point (b, a) in the function's inverse relation, so any x-intercepts of $g(x)$ translate to y-intercepts of its inverse, and any y-intercepts of $g(x)$ translate to x-intercepts of its inverse.

$g(0) = 0^3 - 5(0^2) + 3(0) + 9 = 9$

The function $g(x)$ has a y-intercept of $(0, 9)$, so its inverse relation has an x-intercept of $(9, 0)$.

To find the x-intercepts of $g(x)$, we must find the x-values that make $x^3 - 5x^2 + 3x + 9$ equal to 0. So, we must factor this cubic. According to the Rational Root Theorem, any rational roots of this cubic will be in the set $\pm\{1, 3, 9\}$. Use the Remainder Theorem to test these potential roots.

$g(1) = 1^3 - 5(1^2) + 3(1) + 9 = 1 - 5 + 3 + 9 = 8$
$g(-1) = (-1)^3 - 5(-1)^2 + 3(-1) + 9 = -1 - 5 - 3 + 9 = 0$
$g(3) = 3^3 - 5(3^2) + 3(3) + 9 = 27 - 45 + 9 + 9 = 0$

Both -1 and 3 produced function values of 0, so these are both roots of $g(x)$, and $(x + 1)$ and $(x - 3)$ are factors of $x^3 - 5x^2 + 3x + 9$. Multiplied, $(x + 1)(x - 3) = x^2 - 2x - 3$. Divide $x^3 - 5x^2 + 3x + 9$ by $x^2 - 2x - 3$ to find the remaining factor.

$$
\begin{array}{r}
x - 3 \\
x^2 - 2x - 3 \overline{)\ x^3 - 5x^2 + 3x + 9} \\
\underline{-(x^3 - 2x^2 - 3x)} \\
-3x^2 + 6x + 9 \\
\underline{-(-3x^2 + 6x + 9)} \\
0
\end{array}
$$

The third factor is $(x - 3)$, so the fully factored form of $x^3 - 5x^2 + 3x + 9$ is $(x + 1)(x - 3)(x - 3)$. The function $g(x)$ has x-intercepts of $(-1, 0)$ and $(3, 0)$, so its inverse relation has y-intercepts of $(0, -1)$ and $(0, 3)$.

The root 3 is a repeated root, so $g(x)$ has a local maximum or minimum at $(3, 0)$. The polynomial $x^3 - 5x^2 + 3x + 9$ is a cubic with a positive leading coefficient, so the left arm of $g(x)$ points down and the right arm points up. Given the x- and y-intercepts and end behavior, the point $(3, 0)$ is a local minimum of $g(x)$. The graph of $g(x)$ is shown below.

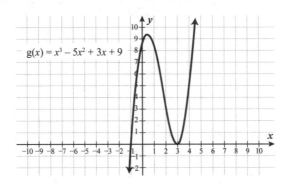

For the inverse relation to be a function, the domain of $g(x)$ must be limited to include a portion of the graph that only increases or only decreases. To find the minimum a-value for which $x \geq a$ describes this domain restriction, we must find the minimum x-value for the last portion of this graph of $g(x)$, where it is only increasing. This point is the local minimum, $(3, 0)$, because to the right of here, $g(x)$ only increases. For $x \geq 3$, $g(x) = x^3 - 5x^2 + 3x + 9$ has an inverse relation that is a cube root function.

4. **see graph on the following page**
    As x approaches $-\infty$, $-4 \cdot 2^x$ approaches 0, so $f(x)$ approaches 16. The graph has a horizontal asymptote of 16 for its left arm. The graph is always decreasing, because of the negative coefficient of the exponential term, so the right arm approaches $-\infty$.

    When $x = 0$, $f(x) = -4 \cdot 2^0 + 16 = 12$, so the y-intercept is 12. The x-intercept occurs when $f(x) = 0$.

$0 = -4 \cdot 2^x + 16$	Substitute 0 for $f(x)$.
$4 \cdot 2^x = 16$	Add $4 \cdot 2^x$ to both sides.
$2^x = 4$	Divide both sides by 4.
$x = \log_2 4$	Take the logarithm, base 2, of both sides.
$x = 2$	Evaluate $\log_2 4$.

So, the x-intercept is 2. Find some more points along the graph, using the function equation.

$f(-2) = -4 \cdot 2^{-2} + 16 = -4 \cdot 1/4 + 16 = 15$
$f(-1) = -4 \cdot 2^{-1} + 16 = -4 \cdot 1/2 + 16 = 14$
$f(1) = -4 \cdot 2^1 + 16 = -4 \cdot 2 + 16 = 8$
$f(3) = -4 \cdot 2^3 + 16 = -4 \cdot 8 + 16 = -16$

So, the graph also passes through (−2, 15), (−1, 14), (1, 8), and (3, −16). The graph of $f(x) = -4 \cdot 2^x + 16$ is shown below.

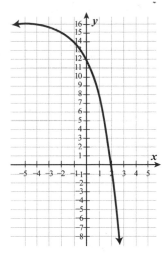

5. **$r(x) = -|3/2\,x + 6| + 3$**

The absolute value function $r(x)$ has a maximum point, so it opens downward, like an upside-down V. This means that the coefficient of the absolute value expression (outside the absolute value expression) must be negative. If the left arm of $r(x)$ has a slope of 3/2, then the right arm has the opposite slope, −3/2.

The maximum value of $r(x)$ is 3. If a downward-opening absolute value function with a maximum of 3 and slopes of 3/2 and −3/2 were centered at the y-axis, it would have an equation of $y = -|3/2\,x| + 3$. However, because $r(x)$ has a maximum at (−4, 3), its line of symmetry is $x = -4$. To translate $y = -|3/2\,x| + 3$ to the left 4 units, we must add 4 to each x-value.

$r(x) = -|3/2\,(x + 4)| + 3$	Replace x with (x + 4).
$r(x) = -|3/2\,x + 6| + 3$	Distribute the multiplication to both terms within the absolute value.

Alternatively, you could use the slope and the maximum point to write the equations of the lines that describe each ray of the absolute value function.

For the left ray, use slope 3/2 and point (−4, 3) in the slope-intercept form of a linear equation.

$y - 3 = 3/2\,(x + 4)$
$y - 3 = 3/2\,x + 6$
$y = 3/2\,x + 9$

For the right ray, use slope −3/2 and point (−4, 3).

$y - 3 = -3/2\,(x + 4)$
$y - 3 = -3/2\,x - 6$
$y = -3/2\,x - 3$

Using the second from last line of each equation-writing process, or the fact that the maximum, 3, occurs when the absolute value expression is equal to 0, we can translate these two equations to the absolute value function $r(x) = -|3/2\, x + 6| + 3$.

6. $-1 - \sqrt{13}$ **and 4**

Algebraically solve for the cases where the argument of the absolute value is non-negative and where it is negative, meaning when $x - 1 \geq 0$ and when $x - 1 < 0$.

When $x \geq 1$, $x - 1 \geq 0$, so $|x - 1| = x - 1$.

$1/2\, x^2 - 5 = x - 1$	Rewrite the equation with the right side equal to $x - 1$.
$1/2\, x^2 - x - 4 = 0$	Subtract x and add 1 to both sides.
$x^2 - 2x - 8 = 0$	Multiply both sides by 2.
$(x - 4)(x + 2) = 0$	Factor the quadratic.
$x = 4, x = -2$	Solve for each factor set equal to 0.

This solution set is for the case where $x \geq 1$, so only $x = 4$ is a solution, while $x = -2$ is an extraneous solution.

When $x < 1$, $x - 1 < 0$, so $|x - 1| = -(x - 1)$, or $-x + 1$.

$1/2\, x^2 - 5 = -x + 1$	Rewrite the equation with the right side equal to $-x + 1$.
$1/2\, x^2 + x - 6 = 0$	Add x and subtract 1 from both sides.
$x^2 + 2x - 12 = 0$	Multiply both sides by 2.
$x = \dfrac{-2 \pm \sqrt{2^2 - 4(1)(-12)}}{2(1)}$	Use the quadratic formula to solve for x.
$x = \dfrac{-2 \pm \sqrt{52}}{2}$	Simplify the radicand.
$x = \dfrac{-2 \pm 2\sqrt{13}}{2}$	Rewrite $\sqrt{52}$ as $\sqrt{4 \cdot 13}$, which simplifies as $2\sqrt{13}$.
$x = -1 \pm \sqrt{13}$	Divide the numerator by the denominator.

This solution set is only for x-values less than 1. The value of $-1 + \sqrt{13}$ is approximately equal to 2.6, so this is an extraneous solution. The only true solution for this portion of the domain is $x = -1 - \sqrt{13}$. The solutions to $1/2\, x^2 - 5 = |x - 1|$ are $x = 4$ and $x = -1 - \sqrt{13}$.

Alternatively, you could graph the functions $y = 1/2\, x^2 - 5$ and $y = |x - 1|$, as shown on the next page. The points of intersection have x-values of 4 and about -4.6 (or $-1 - \sqrt{13}$), so these are the solutions to the given equation.

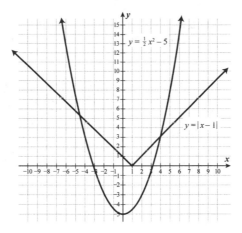

7. **step function**

For the first year, an employee's hourly wage is $12, so for $0 \leq x < 1$, $p(x) = 12$. (An employee can only have worked there a minimum of 0 years, so negative x-values do not make sense in this situation.) On the first anniversary of when he or she was hired (when $x = 1$), the employee's hourly wage increases by 10%.

$12 + 0.1(12) = 13.2$

The employee earns $13.20 per hour for his or her second year of work there. Then, when $x = 2$, the hourly wage increases by 10% again (but 10% of the current amount).

$13.2 + 0.1(13.2) = 14.52$

On the third anniversary of an employee's hire date, he or she gets $14.52 + 0.1(14.52) = 15.972$ dollars per hour. But, we must round off to the nearest cent: $15.97. On the fourth anniversary of the hire date, the employee gets an hourly wage of $15.97 + 0.1(15.97) = 17.567$, which rounds off to $17.57. An employee who was hired 4 1/2 years ago is still earning $17.57 per hour, because the amount will not change again until his or her fifth anniversary of working there. So, the function $p(x)$ can be defined as follows.

$$p(x) = \begin{cases} 12.00, & 0 \leq x < 1 \\ 13.20, & 1 \leq x < 2 \\ 14.52, & 2 \leq x < 3 \\ 15.97, & 3 \leq x < 4 \\ 17.57, & 4 \leq x \leq 4.5 \end{cases}$$

Each section of the graph of $p(x)$ is a horizontal line segment, so this is a step function.

8. see graph below

Along the flat stretch of road, Naima rode at a constant speed of 0.2 mile per minute, so her distance traveled was equal to 0.2x. To write a piecewise-defined function, we need to know at what x-value, or after how many minutes, she hit the downhill. We know she traveled 1.2 miles in this time.

$1.2 = 0.2x$	Set distance traveled, 1.2, equal to 0.2x.
$6 = x$	Divide both sides by 0.2.

So, for $0 \leq x \leq 6$, Naima's distance traveled was given by the function $s(x) = 0.2x$. After 6 minutes, her distance traveled downhill was given by $d = 0.1m^2 + 0.4m$, where m was the number of minutes since she started going downhill. Notice that m is different from x, the number of minutes since Naima began her ride. Because $m = 0$ when $x = 6$, and all x-values will be 6 minutes more than all m-values, we can say that $x = m + 6$, or $m = x - 6$. Also, d is different from $s(x)$. When $d = 0$, $s(x) = 1.2$, and all s(x)-values will be 1.2 miles greater than all d-values, so $s(x) = d + 1.2$.

$s(x) = 0.1m^2 + 0.4m + 1.2$	Substitute $(0.1m^2 + 0.4m)$ for d.
$s(x) = 0.1(x - 6)^2 + 0.4(x - 6) + 1.2$	Substitute $(x - 6)$ for m.
$s(x) = 0.1(x^2 - 12x + 36) + 0.4(x - 6) + 1.2$	Expand the squared binomial.
$s(x) = 0.1x^2 - 1.2x + 3.6 + 0.4x - 2.4 + 1.2$	Use the distributive property.
$s(x) = 0.1x^2 - 0.8x + 2.4$	Combine all like terms.

This equation describes $s(x)$ only for times 6 minutes or more into Naima's ride. Her ride was a total of 8 minutes long. So, her total distance traveled throughout this ride is given by the piecewise-defined function below.

$$s(x) = \begin{cases} 0.2x, \, 0 \leq x \leq 6 \\ 0.1x^2 - 0.8x + 2.4, \, 6 < x \leq 8 \end{cases}$$

To graph this, we draw a straight line segment along $y = 0.2x$ from (0, 0) to (6, 1.2) and draw the quadratic using points along it, starting at (6, 1.2). (Notice that $0.1(6^2) - 0.8(6) + 2.4$ is also equal to 1.2, so the function is continuous.)

$s(7) = 0.1(7^2) - 0.8(7) + 2.4 = 4.9 - 5.6 + 2.4 = 1.7$
$s(8) = 0.1(8^2) - 0.8(8) + 2.4 = 6.4 - 6.4 + 2.4 = 2.4$

The quadratic goes from (6, 1.2) through (7, 1.7) and to (8, 2.4). The graph of $s(x)$ is shown below.

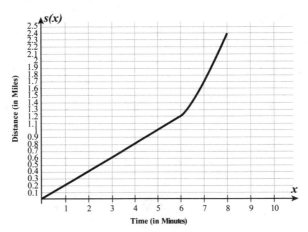

9. **173 years**

    If 77% of the original sample quantity remains, then the ratio of the remaining quantity to the original quantity (or N/A) is equal to 0.77.

$N/A = e^{kt}$	Divide both sides of $N = A \cdot e^{kt}$ by A, to get N/A on the left side.
$0.77 = e^{65k}$	Substitute 0.77 for N/A and 65 for t.
$\ln 0.77 = 65k$	Take the ln of both sides.
$\dfrac{\ln 0.77}{65} = k$	Divide both sides by 65.
$k \approx -0.004$	Use your calculator to find the value of $\dfrac{\ln 0.77}{65}$ and round off.

    So, the exponential decay of silicon-32 is given by the equation $N = A \cdot e^{-0.004t}$. This can also be written as $N/A = e^{-0.004t}$. To find the half-life of silicon-32, substitute 1/2, or 0.5, for N/A.

$0.5 = e^{-0.004t}$	
$\ln 0.5 = -0.004t$	Take the ln of both sides.
$\dfrac{\ln 0.5}{-0.004} = t$	Divide both sides by −0.004.
$t \approx 173$	Use your calculator to find the value of $\dfrac{\ln 0.5}{-0.004}$ and round off.

    The half-life of silicon-32 is about 173 years.

REFLECT

**Congratulations on completing Chapter 6!
Here's what we just covered.
Rate your confidence in your ability to**

- Graph cube root functions

  ① ② ③ ④ ⑤

- Graph piecewise-defined functions, including step functions (such as floor and ceiling functions) and absolute value functions

  ① ② ③ ④ ⑤

- Solve single-variable absolute value equations, both algebraically and by graphing systems of equations

  ① ② ③ ④ ⑤

- Graph exponential functions

  ① ② ③ ④ ⑤

- Write and use cube root, piecewise-defined, and exponential functions to represent real-life situations and solve problems

  ① ② ③ ④ ⑤

If you rated any of these topics lower than you'd like, consider reviewing the corresponding lesson before moving on, especially if you found yourself unable to correctly answer one of the related end-of-chapter questions.

Access your online student tools for a handy, printable list of Key Points for this chapter. These can be helpful for retaining what you've learned as you continue to explore these topics.

Chapter 7
Making and Using
Mathematical Models

GOALS By the end of this chapter,
you will be able to:

- Manipulate given equations to solve for certain variables or to reveal properties of functions

- Determine key features of functions from equations, tables, and graphs, and interpret those features in the context of the situations described by the functions

- Model described situations with equations and inequalities, and use them to solve problems

- Model described situations with graphs, and use them to solve problems

Lesson 7.1
Manipulating Equations

Many relationships in the world around us can be modeled using math, in the form of equations, inequalities, and graphs. When we are given a representative equation, we may be able to manipulate it to make better use of it.

For example, if we have the formula that converts temperatures in degrees Celsius (variable C) to degrees Fahrenheit (variable F), $F = 9/5\ C + 32$, then we can rewrite this equation to perform conversions in the other direction. Solve for C, using the same algebraic solution methods that you would use to solve a single-variable equation, treating F as just another number.

$F = 9/5\ C + 32$
$F - 32 = 9/5\ C$ Subtract 32 from both sides.
$5/9\ (F - 32) = C$ Multiply both sides by 5/9, the reciprocal of 9/5.

Now we have the formula $C = 5/9\ (F - 32)$, which can be used to directly convert temperatures in degrees Fahrenheit to degrees Celsius.

Here is how you may see manipulation of equations on the ACT.

Oliver discovers that the current value of his car can be calculated using the equation $x = P(1 - d)^n$, where x is the current value, P is the original price Oliver paid for the car, d is a depreciation constant for that kind of car, and n is the number of years since he bought it. Which of the following is an expression for d in terms of P, n, and x?

A. $1 - \sqrt[n]{\dfrac{x}{P}}$

B. $\sqrt[n]{\dfrac{x+1}{P}}$

C. $\sqrt[n]{x - P}$

D. $\sqrt[n]{1 - \dfrac{x}{P}}$

E. $\dfrac{\sqrt[n]{x}}{P} + 1$

EXAMPLE 1

If a car accelerates at a steady rate, then the distance, d, the car travels in time, t, is given by the formula $d = v_i t + 1/2\, at^2$, where v_i is the car's initial velocity and a is its acceleration. Write a formula that gives a in terms of v_i and t. Then, use the fact that the rate of acceleration is the ratio of the change in velocity (from the car's initial velocity to its final velocity, v_f) to the change in time, to write a formula for v_f in terms of d, t, and v_i.

First, we must solve the given equation for a.

$d = v_i t + 1/2\, at^2$

$d - v_i t = 1/2\, at^2$ Subtract $v_i t$ from both sides.

$2(d - v_i t) = at^2$ Multiply both sides by 2.

$\dfrac{2(d - v_i t)}{t^2} = a$ Divide both sides by t^2.

The formula for acceleration is $a = \dfrac{2(d - v_i t)}{t^2}$, which can also be written as $a = \dfrac{2d}{t^2} - \dfrac{2v_i}{t}$.

The fraction $\dfrac{2(d - v_i t)}{t^2}$ can be written as $\dfrac{2d - 2v_i t}{t^2}$ and then broken into the sum of two fractions with the same denominator: $\dfrac{2d}{t^2} - \dfrac{2v_i t}{t^2}$. By canceling out a factor of t in the numerator and denominator of the second fraction, we get $\dfrac{2d}{t^2} - \dfrac{2v_i}{t}$.

The acceleration rate is the change in velocity over the change in time. The variable t represents the time that the car has been moving distance d, so the change in time is just t. The change in velocity is $v_f - v_i$.

$a = \dfrac{v_f - v_i}{t}$

Let's substitute this expression for a into our formula.

$\dfrac{v_f - v_i}{t} = \dfrac{2d}{t^2} - \dfrac{2v_i}{t}$

Now we can solve for v_f.

$v_f - v_i = t\left(\dfrac{2d}{t^2} - \dfrac{2v_i}{t}\right)$ Multiply both sides by t.

$v_f - v_i = \dfrac{2d}{t} - 2v_i$ Distribute the factor of t to both terms in the parentheses.

$v_f = \dfrac{2d}{t} - v_i$ Add v_i to both sides.

This formula gives v_f in terms of d, t, and v_i.

Sometimes, instead of solving for a different variable, you just need to rearrange a given equation to reveal various properties of the function. For example, in Chapter 1, Example 13, we rearranged the given function $h = -16t^2 + 48t + 64$ in factored form as $h = = -16(t + 1)(t - 4)$ to find its x-intercepts (t-intercepts), -1 and 4, and in vertex form as $h = -16(t - 3/2)^2 + 100$ to find its vertex, $(3/2, 100)$.

EXAMPLE 2

To graph the function $y = \dfrac{3x^2 - 12}{x^2 + x - 6}$, Bonnie wants to determine the horizontal asymptote, the vertical asymptote, and any holes in the graph. Which of these can she best assess from the current form of the function? Rewrite the function to reveal the other two properties.

When the numerator and denominator of a rational function are of the same degree, as they are here, the horizontal asymptote is the line

$y = \dfrac{\text{leading coefficient of numerator}}{\text{leading coefficient of denominator}}$. So, in this form, Bonnie can easily see that

the function has a horizontal asymptote at $y = 3/1$, or $y = 3$.

To determine the vertical asymptote and any holes in the function, she must factor both the numerator and the denominator.

$y = \dfrac{3\left(x^2 - 4\right)}{x^2 + x - 6}$ 　　　　Factor 3 out of the numerator.

$y = \dfrac{3\left(x + 2\right)\left(x - 2\right)}{\left(x + 3\right)\left(x - 2\right)}$ 　　　Factor the quadratics in the numerator and denominator.

The factored form of the rational function also tells you that the function has an *x*-intercept of –2, because that value will make the factor (*x* + 2) of the numerator equal to 0 (but will not make the denominator 0).

Any zero of the denominator represents a point of discontinuity in the function. If the zero is also a zero of the numerator, then the discontinuity is a hole. If the zero is not a zero of the numerator, then the discontinuity is a vertical asymptote. This function has a hole when $x = 2$ and a vertical asymptote of $x = -3$.

$x = P(1 - d)^n$

$x/P = (1 - d)^n$ 　　　　Divide both sides by *P*.

$\sqrt[n]{\dfrac{x}{P}} = 1 - d$ 　　　　Take the *n*th root of both sides.

$\sqrt[n]{\dfrac{x}{P}} - 1 = -d$ 　　　　Subtract 1 from both sides.

$-\sqrt[n]{\dfrac{x}{P}} + 1 = d$ 　　　　Multiply both sides by –1.

$1 - \sqrt[n]{\dfrac{x}{P}} = d$ 　　　　Use the commutative property of addition.

The correct answer is (A).

Lesson 7.2
Interpreting Equations, Tables, and Graphs

From equations, tables, and graphs, you can determine various key features of functions, such as intercepts, periods of increase or decrease, and much more. Interpreting these features within the context of the situation that the function describes reveals information about the relationship it represents.

EXAMPLE 3

Felice bought shares in a consumer electronics and computer software company, and has been tracking the price per share on the stock market for the past 8 days, as shown in the graph below.

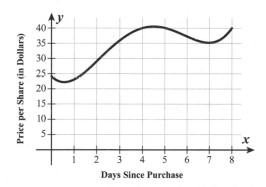

Between which days (whole numbers only) was the share price increasing? What does the *y*-intercept of this graph represent in the context of the situation? By approximately how much did the share price increase from when Felice bought it to day 8? Between which whole numbers of days did the share price reach its maximum?

The graph rises from left to right between $x = 1$ and $x = 4$ and again between $x = 7$ and $x = 8$. The share price was increasing between day 1 and day 4, as well as between day 7 and day 8.

The y-intercept is where the graph intersects the y-axis, when $x = 0$. The x-values represent days since Felice purchased the shares, and the y-values represent share prices, so the y-intercept at (0, 24) represents a share price of $24 when Felice purchased the shares. She bought the shares for $24 each.

On day 8, the graph indicates a share price of $40. From when Felice bought the shares to day 8, the share price increased by $16 (40 − 24 = 16).

The share price reached its maximum, above $40, between 4 and 5 days after Felice bought the shares.

EXAMPLE

In Chapter 6, Lesson 6.2, we explored absolute value functions. Here is a real-life application for a situation where only the positive difference in length is wanted. Without an absolute value sign, the function would produce negative values for any measurement less than 189.4 and positive values for any measurement greater than 189.4.

This function must represent how far off the students' measurements are from the actual length of the desk.

Mr. Davis asked each of his students to measure the length of his desk to the nearest tenth of a centimeter, then recorded their measurements. He said that the relative error of each of their measurements is a function value on the graph of $r(x) = \dfrac{|x - 189.4|}{189.4}$. Relative error of a measurement is the ratio of the absolute error of the measurement to the true value (in this case the actual length of the desk). What do x and $r(x)$ represent in this function? What does the number 189.4 represent? What are the limitations on the domain of this function, given the situation? What does the x-intercept of the graph represent?

Relative error is defined to be a ratio of absolute error of the measurement to the actual (true) measurement, so $r(x)$ represents the relative error, and 189.4 is the actual measurement. Mr. Davis's desk is actually 189.4 centimeters long. So, x must represent a measurement result, such as those of Mr. Davis's students.

Because x is a measurement of length, it must be a positive number. The domain of $r(x)$ is $x > 0$.

The x-intercept occurs when $r(x) = 0$, so the x-intercept represents a measurement of exactly 189.4, resulting in a relative error of 0.

One day, Mariah recorded the outdoor temperature in Nome, Alaska, at various times throughout the day. She recorded the values in the table below, with x representing hours into the day (for example, $x = 6$ represents 6:00 A.M.) and $f(x)$ representing temperature, in degrees Fahrenheit.

x	$f(x)$
6	–5
8	3
10	12
11	20
14	29
16	23
19	4

If the temperatures were continuously recorded throughout the day, the graph of $f(x)$ could be drawn as a continuous function with a domain of all real numbers from $x = 0$ to $x = 24$. In what range of times must there be an x-intercept for this function? What does this represent? If there is only one local maximum value for the function, in what range of times must it be? If the function has no local minimums, for what period of time is the function definitely decreasing?

An x-intercept occurs when $f(x) = 0$. Because the function is continuous and has a negative value at $x = 6$ and a positive value at $x = 8$, it must cross the x-axis somewhere between 6 and 8. So, the temperature must have been 0°F sometime between 6:00 A.M. and 8:00 A.M.

Although the maximum $f(x)$-value given in Mariah's table is 29, this may or may not be the maximum value for the function. We know that the temperature was 20° F at 11:00 A.M. and 29° F at 2:00 P.M., but we do not know the temperatures in between those times. Perhaps the temperature reached a maximum higher than 29° F and was falling again by 2:00 P.M. Or, it's possible that the maximum didn't occur until after 2:00 P.M. but before 4:00 P.M., when the temperature was down to 23° F. All we can conclude is that the local maximum temperature must have occurred sometime between 11:00 A.M. and 4:00 P.M.

If the function has only one local maximum and no local minimums, then it only decreases after its maximum. We do not know exactly when the maximum occurred, but we know it was before 4:00 P.M. After 4:00 P.M., the temperature is definitely decreasing (from 4:00 P.M. to midnight).

With only one local maximum for the function, the temperatures only change from increasing to decreasing once during this day. If there were multiple local maximums, then there could have been another local maximum, for example at 6:00 P.M., with a higher temperature than the one earlier in the day.

EXAMPLE 6

Toby used the formulas for the surface area of a sphere and the volume of a sphere to find formulas for the radius of a sphere, given its surface area or volume. Using his calculator, he simplified the coefficients in the two functions, to get $r = 0.28\sqrt{S}$ and $r = 0.62\sqrt[3]{V}$. These two functions are shown on the graphs below.

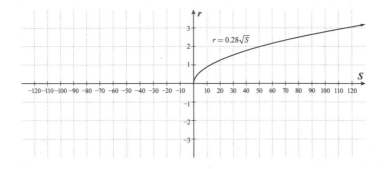

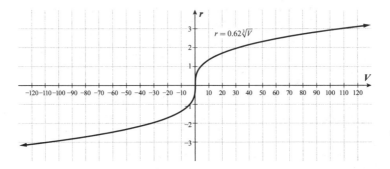

What is the difference between the domains of the two graphed functions? Why are they different? What additional domain restriction(s), if any, should Toby use to make the graphs appropriately represent the data for this particular situation?

The graph of $r = 0.62\sqrt[3]{V}$ has a domain of all real numbers, because its arms are extending forever to the left and right, as a cube root function. The graph of $r = 0.28\sqrt{S}$, however, has a domain of only $S \geq 0$. It begins at the r-axis and continues forever to the right. The square root $\sqrt{S}$ is always positive, so the function $r = 0.28\sqrt{S}$ only has positive function values, unlike the cube root function, which can have both positive and negative function values.

You could also argue that a sphere with a surface area of 0 or a volume of 0 does not exist as a sphere, which means that the domain for $r = 0.28\sqrt{S}$ should be restricted to $S > 0$ and the domain for $r = 0.62\sqrt[3]{V}$ should be restricted to $V > 0$.

The domain of the function $r = 0.28\sqrt{S}$ represents surface areas of spheres, so $S \geq 0$ is an appropriate domain. The domain of function $r = 0.62\sqrt[3]{V}$ represents volumes of spheres, so negative values of V are inappropriate. Toby should restrict the domain of $r = 0.62\sqrt[3]{V}$ to $V \geq 0$ for this particular situation.

6

Emily and Masaru decided to start a glee club at their school and were the first two members. Values for the total number of students in the glee club, $g(x)$, after x weeks are shown in the table below.

x	0	1	3	4	5	7	9	14	16
$g(x)$	2	10	16	18	20	23	26	32	34

What does the y-intercept (the $g(x)$-intercept) of this function represent? What was the average rate of change of $g(x)$ from $x = 0$ to $x = 16$? What does this represent? What was the average rate of change of $g(x)$ from $x = 0$ to $x = 4$? What does this represent? Based on this information, what can you conclude about the growth of the glee club over the time period shown?

The x-values represent weeks since the glee club was founded, and the $g(x)$-values represent the total number of glee club members at that time. The y-intercept, when $x = 0$, represents the glee club membership after 0 weeks, when it had just been founded. According to the table, when $x = 0$, $g(x) = 2$, so at founding, the glee club had exactly 2 members.

When $x = 16$, there are 34 glee club members. The rate of change of club membership is the change in membership divided by the change in time (in this case, the number of weeks since the club was founded).

$$\frac{34 - 2}{16 - 0} = 32/16 = 2$$

The rate of change from $x = 0$ to $x = 16$ is 2. This means that, over the first 16 weeks, the glee club's membership increased at an average rate of 2 members per week.

Look at the table to find the x-value of 4, which corresponds to a $g(x)$-value of 18. Find the average rate of change from (0, 2) to (4, 18).

$$\frac{18 - 2}{4 - 0} = 16/4 = 4$$

The rate of change from $x = 0$ to $x = 4$ is 4. This means that, over the first 4 weeks, the glee club's membership increased at an average rate of 4 members per week. This rate of change is twice the rate of change for the entire 16 weeks shown, even though 4 weeks is only 1/4 of the full time period. The glee club's membership increased very quickly over the first 4 weeks but slowed down substantially over later weeks.

EXAMPLE 8

A town has two different postal stations, one for the east side and one for the west side, with the same number of households served by each station. All the mail carriers working within a station have the same average mail delivery rate, in households per hour, but the east-side mail carriers have a different delivery rate than the west-side mail carriers. Examples of time, t, in hours, that it takes x east-side mail carriers, working simultaneously, to deliver to all east-side households are given in the table below.

x	t
3	40
4	30
6	20
8	15
15	8

Here is how you may see interpreting graphs on the SAT.

After Hurricane Katrina caused high rates of homelessness in New Orleans, efforts were made to provide housing opportunities for the homeless. Estimates of the homeless population are shown in the scatterplot below.

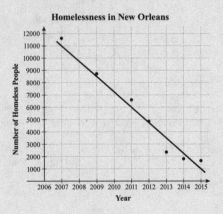

Based on the line of best fit in the scatterplot above, which of the following is closest to the average annual decrease in the homeless population in New Orleans between 2007 and 2015?

A) −1000

B) −1250

C) −1500

D) −1750

On the west side, the number of mail carriers working simultaneously and the number of hours it takes them to deliver to all west-side households are inversely proportional, with a constant of variation of 128. On which side of town do the mail carriers have a faster mail delivery rate? What kind of function can be written that correctly relates *t* to *x* for each side of town? For each of these functions, based on the domain values for the given situation, is a graph of the function or a table of values a more appropriate model for the relationship?

In the table of values for east-side carriers, when *x* doubles, *t* is halved, so *t* varies inversely as *x*. This means that the product of *t* and *x* is the constant of proportion, which in this case is 120 ($3 \cdot 40 = 120$, $4 \cdot 30 = 120$, and so on). So, $t = 120/x$ describes the relationship in this table.

The number of west-side mail carriers and the time it takes them to deliver all west-side mail is again an inverse variation relationship, this time with a constant of variation of 128. So, $t = 128/x$ describes the relationship for the west-side mail carriers.

For any positive *x*-value, the equation $t = 128/x$ will produce a greater *t*-value than the equation $t = 120/x$. The variable *t* represents the time it takes the mail to be delivered, so it takes the west-side mail carriers more time to deliver the mail. The same number of households is served by each station, so the west-side mail carriers are slower. The east-side mail carriers have a faster delivery rate.

The functions $t = 128/x$ and $t = 120/x$ are both rational functions. The domain, or set of *x*-values, for each function represents the number of mail carriers simultaneously working to deliver mail on one side of town. So, the domain consists only of whole numbers for this situation. A graph of either rational function would include all real-number values of *x* (except for 0), including all values between consecutive whole numbers. A table of values is a more appropriate model, because *x*-values can be limited to only whole numbers in the table.

Compare the values in two rows of the table. When $x = 3$, $t = 40$, and when $x = 6$, $t = 20$. In this case, doubling *x* resulted in halving *t*. Compare another two rows in the table to make sure the same relationship holds true. When $x = 4$, $t = 30$, and when $x = 8$, $t = 15$. Again, doubling *x* caused *t* to be divided by 2. The relationship is an inverse variation.

EXAMPLE 9

The cross-section of a satellite dish is in the shape of a portion of a parabolic curve. Luis graphed the cross-section of his television satellite dish on a coordinate grid, with the rim of the dish aligned with the *x*-axis and the center of the dish aligned with the *y*-axis, as shown below. Units on both axes represent distances in inches.

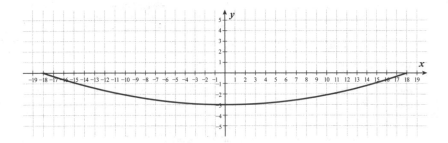

Priscilla has a television satellite dish with a cross-section that can be similarly graphed using the equation $y = 1/36\ x^2 - 9/4$, with $y \leq 0$. Which function has a greater positive *x*-intercept? What does this value represent? Which function has a minimum with a greater absolute value? What does this value represent? How do the dimensions of Luis's and Priscilla's satellite dishes compare?

The *x*-intercept of Luis's graph is 18. The *x*-intercept occurs when $y = 0$, so substitute 0 for *y* in the equation for Priscilla's satellite dish.

$0 = 1/36\ x^2 - 9/4$

$9/4 = 1/36\ x^2$	Add 9/4 to both sides.
$81 = x^2$	Multiply both sides by 36.
$\pm 9 = x$	Take the square root of both sides.

The slope of a line is the change in *y*-values over the change in *x*-values. In this case, the slope represents a change in the number of homeless people per year, as an average.

Use the points on the line of best fit for the years 2007 and 2015. The line of best fit passes through the points (2007, 11000) and (2015, 1000).

$$\text{slope} = \frac{1000 - 11000}{2015 - 2007} = -10000/8 = -1250$$

The slope of the line is −1250, which represents a decrease (because it is negative) in the number of homeless people in New Orleans, of about 1250 people per year. The correct answer is (B).

Priscilla's function has *x*-intercepts of −9 and 9. Comparing the positive intercepts, Luis's *x*-intercept of 18 is greater than Priscilla's *x*-intercept of 9. Both functions are symmetrical with respect to the *y*-axis, so the positive *x*-intercept value represents the radius of the satellite dish.

The minimum function value on Luis's graph is −3. The minimum function value of the quadratic function $y = 1/36\ x^2 - 9/4$ is −9/4, which is equal to −2 1/4. Luis's function has a minimum with a greater absolute value, 3, as compared to Priscilla's, 2 1/4. This represents the depth, or height, of each satellite dish.

The diameter of a circle is twice the radius. Luis's satellite dish is 36 inches in diameter and 3 inches deep, and Priscilla's satellite dish is 18 inches in diameter and 2 1/4 inches deep. Luis's satellite dish is twice as wide as Priscilla's but just a little deeper.

The graph of Priscilla's satellite dish is defined to be limited to $y \leq 0$, as in Luis's graph. Another way to view the limitation is in terms of the domain of the graph. Luis's graph has a domain of $-18 \leq x \leq 18$, or $|x| \leq 18$, and Priscilla's graph has a domain of $-9 \leq x \leq 9$, or $|x| \leq 9$. This shows that the parabolic curves each extend only to a maximum of the radius length in each direction from center.

Lesson 7.3
Modeling Situations With Equations and Inequalities

EQUATIONS AND INEQUALITIES IN ONE VARIABLE

EXAMPLE

A family went out to eat at a restaurant, where they used a coupon that gave them a 20% discount on food (but not drinks), before tax. They paid a total of $50.40, including 8% tax and an 18% tip (both percents based on the pre-tax total). If their bill included $12 for drinks, how much did they save by using the coupon?

You can also write the total amount of their discounted food order as the full-price amount minus 20% of that amount: $f - 0.2f$. This also simplifies to $0.8f$.

We must write an equation that models this situation. Although we want to solve for the family's savings in the end, the basic unknown information is how much the food portion of their order cost. Let f = the price of the food they ordered. With a 20% discount on food, they only paid 80% of the food's price, or $0.8f$.

The family also ordered $12 in drinks, so the total price of their order was $0.8f + 12$. They also paid 8% tax and 18% tip on this amount. We can write the total amount they paid as $(0.8f + 12) + 0.08(0.8f + 12) + 0.18(0.8f + 12)$, which simplifies, using the distributive property, as $(1 + 0.08 + 0.18)(0.8f + 12)$, or $1.26(0.8f + 12)$. Set this expression equal to the total amount the family paid.

Alternatively, you could first distribute through to get $1.008f + 15.12 = 50.4$, then subtract 15.12 from both sides to get $1.008f = 35.28$, and finally divide both sides by 1.008 to get $f = 35$.

$$1.26(0.8f + 12) = 50.4$$

We can now solve the equation for f.

$0.8f + 12 = 40$	Divide both sides by 1.26.
$0.8f = 28$	Subtract 12 from both sides.
$f = 35$	Divide both sides by 0.8.

With the original price of the family's food order, we can now calculate how much they would have paid without the coupon. The total price of their food and drinks would be $35 + 12 = 47$ dollars. They would still have paid an additional 8% tax and 18% tip, for a total extra 26% of the pre-tax total.

Alternatively, we could calculate the savings by finding 20% of the total food price, $35, and adding tax and tip on that amount. 20% of $35 is $7. The total of $7, 8% of $7, and 18% of $7 is $8.82.

$$1.26(47) = 59.22$$

Without the coupon, the family would have paid $59.22 total. Subtract the amount they actually paid.

$$59.22 - 50.40 = 8.82$$

The family saved a total of $8.82 by using the coupon.

10

EXAMPLE 11

In Chapter 6, Lesson 6.3, we learned that real-life exponential growth functions are typically written in the form $f(x) = a \cdot b^x$, with $b > 1$. Here, we must write an exponential expression to represent the population growth but then also use the expression in an inequality to solve for when its value is greater than one million.

In Sam's hometown, the population has been increasing at an annual rate of 2.5% since 2010, when the population was 790,390. If the population continues to increase at this rate, in what year will the population be greater than one million?

Because the increase in population is a percent of each previous year's population, we must use an exponential expression to represent Sam's hometown population at a given time. Let t represent years since 2010 (so $t = 0$ represents the year 2010). The city's population is multiplied by 1.025 each year. (The 1 represents the same population number repeated, and the 0.025 represents the additional 2.5% of that number in growth.) So, the population in 2010 is multiplied by 1.025 t times to find the population after t years: $790{,}390 \cdot 1.025^t$.

Let's set up an inequality to find when this population is greater than one million (1,000,000).

$$790,390 \cdot 1.025^t > 1,000,000$$

$1.025^t > 1.265198$	Divide both sides by 790,390. (Round off the quotient.)
$t > \log_{1.025} 1.265198$	Take the logarithm, base 1.025, of both sides.
$t > \dfrac{\log 1.265198}{\log 1.025}$	Use the change of base property to rewrite with common logarithms.
$t > 9.526$	Use your calculator to find the approximate value of the ratio.

Sam's hometown will have a population of over 1,000,000 about 9.526 years after 2010. We must round up to the next whole-number year. The population will be greater than 1,000,000 in the year 2020.

EQUATIONS AND INEQUALITIES IN TWO VARIABLES

An equation or inequality in two variables models the relationship between two different sets of values. In many math problems, you are given a specific value for one of the variables and then must use the equation or inequality to solve for the other. Other times, you may need only to express the relationship as an equation or inequality, or as a system of equations or inequalities.

EXAMPLE 12

For a bike ride at a constant speed, on flat ground with no wind, Hiroshi found that his energy output for the ride was related to the speed at which he rode. For every additional mile per hour (mph) in his speed, he used an additional 20% calories per hour. When he rides at 10 mph for an entire bike ride, he burns 124 calories per hour. Write a function that relates the calories per hour, $c(x)$, that Hiroshi burns on a bike ride, to his velocity, x, in miles per hour.

Hiroshi's base daily caloric need is 2500 calories, for a day when he does not ride his bike. Write a function to represent Hiroshi's total caloric needs, $d(x)$, for a day in which he took a 2-hour bike ride at a speed of x miles per hour. Use this function to find the total calories he needs for a day in which he rides for 2 hours at 15 mph.

The first function relates calories burned per hour to speed in mph. For each increase of 1 mph, Hiroshi uses an additional 20% calories per hour. So, this is an exponential relationship, with some original value multiplied by 1.2 for each increase of 1 mph. The base is 1.2, and the exponent is x, the speed in mph. Use the given data to solve for the unknown original value, c_0. Write an exponential function with the original value, c_0, multiplied by 1.2^x to get $c(x)$.

$c(x) = c_0 \cdot 1.2^x$

$124 = c_0 \cdot 1.2^{10}$ Substitute 124 for total calories burned and 10 for velocity.

$124 \approx c_0 \cdot 6.19$ Evaluate 1.2^{10}.

$20.03 \approx c_0$ Divide both sides by 6.19.

Let's round c_0 off to 20. The function relating calories per hour, $c(x)$, that Hiroshi burns on a bike ride, to his velocity, x, in mph, is $c(x) = 20 \cdot 1.2^x$.

Now we want to create a function that relates the total number of calories needed one day to the speed at which Hiroshi rode for a 2-hour bike ride that day. We can use the function relating calories per hour to speed, to find the total calories used for a 2-hour bike ride. Multiply the calories per hour by the number of hours: $2 \cdot (20 \cdot 1.2^x)$, or $40 \cdot 1.2^x$. Additionally, Hiroshi has a base daily caloric need of 2500 calories. So, for a day in which he takes a 2-hour bike ride at a velocity of x mph, he needs a total of $d(x)$ calories, given by the function $d(x) = 40 \cdot 1.2^x + 2500$.

On a day in which Hiroshi rode for 2 hours at 15 mph, $x = 15$. Substitute this value into $d(x) = 40 \cdot 1.2^x + 2500$.

$d(x) = 40 \cdot 1.2^{15} + 2500$
$d(x) \approx 40 \cdot 15.407 + 2500$ Evaluate 1.2^{15}.
$d(x) \approx 616.28 + 2500$ Multiply 40 and 15.407.
$d(x) \approx 3116.28$ Add.

Hiroshi needs about 3116 calories for this day.

Here is how you may see modeling with equations on the SAT.

An aquarium is putting in a new 180-gallon tank to display fish and plants that thrive in a salinity of 1.5 ppt (parts per thousand). An aquarium employee will mix seawater having a salinity of 35 ppt with freshwater having a salinity of 0.5 ppt. Solving which of the following systems of equations yields the number of gallons of seawater, s, and the number of gallons of freshwater, f, that will be combined to fill the tank?

A) $s + f = 180$
    $0.035s + 0.0005f = 0.27$

B) $s + f = 180$
    $(0.035 + 0.0005)(s + f) = 0.0015$

C) $s + f = 180$
    $0.035s + 0.0005f = 0.0015$

D) $s + f = 0.0015$
    $0.035s + 0.0005f = 180$

Lesson 7.4
Modeling Situations with Graphs

Graphs may help you to better understand or express the relationships between variables in a function. Many times you will need to translate information in a problem to a function or set of functions before graphing

In Chapter 6, Lesson 6.2, we looked at ceiling functions and their graphs in Examples 9 and 10. Here, we are translating given information about Rachel's babysitting charges into a ceiling function, which we will then graph.

EXAMPLE 13

Rachel babysits for a maximum of 10 hours per week. She charges $15 per hour or any part of an hour in her total for the week for each client. If one of her clients hires her for more than 6 hours in one week, she gives them a 30% discount off her usual hourly rate, for all hours over 6 hours. Write and graph a function representing the amount, $a(x)$, a client pays Rachel for a week in which she works x hours for them.

Each part of an hour is rounded up to the next whole hour, so this is a ceiling function.

For up to 6 hours, Rachel charges $15 per hour or part of an hour, so for $0 \leq x \leq 6$, $a(x) = 15 \cdot \lceil x \rceil$. For more than 6 hours, Rachel charges $15 for each of the first 6 hours and 70% of $15 per hour for each additional hour. (A 30% discount means paying 70% of the regular rate.) To show the discounted charge for each additional hour, you must subtract the first 6 hours from x before multiplying by the hourly rate. So, for $6 < x \leq 10$, $a(x) = 15(6) + 0.7(15) \cdot \lceil x - 6 \rceil$. This equation simplifies to $a(x) = 90 + 10.5 \cdot \lceil x - 6 \rceil$.

The amount, $a(x)$, a client pays Rachel for a week in which she works x hours for them is given by the piecewise function below.

$$a(x) = \begin{cases} 15 \cdot \lceil x \rceil, \, 0 \leq x \leq 6 \\ 10.5 \cdot \lceil x - 6 \rceil + 90, \, 6 < x < 10 \end{cases}$$

Substitute various values for x to determine the levels of the steps. For example:

$a(1) = 15 \cdot \lceil 1 \rceil = 15 \cdot 1 = 15$
$a(2.5) = 15 \cdot \lceil 2.5 \rceil = 15 \cdot 3 = 45$
$a(6) = 15 \cdot \lceil 6 \rceil = 15 \cdot 6 = 90$
$a(7) = 10.5 \cdot \lceil 7 - 6 \rceil + 90 = 10.5 \cdot \lceil 1 \rceil + 90 = 10.5 \cdot 1 + 90 = 100.5$
$a(9.5) = 10.5 \cdot \lceil 9.5 - 6 \rceil + 90 = 10.5 \cdot \lceil 3.5 \rceil + 90 = 10.5 \cdot 4 + 90 = 132$

Only a number of hours (x or $x - 6$) goes inside the ceiling symbol, because it is the number of hours that is rounded up to the next whole number of hours, not the charge amount rounding up to the next whole dollar.

Here is a graph of $a(x)$.

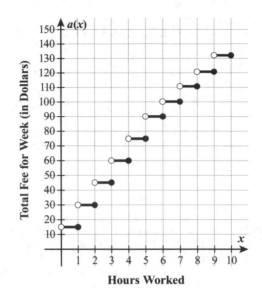

For an example of how graph-modeling questions appear on the ACT, please access the online Student Tools for this book.

The total number of gallons in the tank will be 180, so the sum of the number of gallons of seawater and the number of gallons of freshwater must equal 180: $s + f = 180$.

A salinity of 1.5 parts per thousand can be written as 1.5/1000, or 0.0015. The salt content of the 180-gallon mix will be 0.0015 times 180. With a salinity of 35 ppt, the s gallons of seawater have a salt content of $0.035s$. With a salinity of 0.5 ppt, the f gallons of freshwater have a salt content of $0.0005f$. The sum of the amounts of salt in the saltwater and freshwater equals the amount of salt in the mixed waters.

$0.035s + 0.0005f = 0.0015(180)$
$0.035s + 0.0005f = 0.27$

So, the correct system of equations to use for modeling and solving this problem is the one shown in (A).

An eccentric rich philanthropist, Mrs. Z, held a contest, with the winning organization receiving their choice of the following: A) $10,000 per year for the next 50 years, or B) an account with a starting balance of $5, the contents of which Mrs. Z would double each year, for a maximum of the next 20 years. (The organization could choose to empty the account at any point during those 20 years but could not otherwise deposit to or withdraw from the account during that time.) After about how many years do the two prize options have the same value? Which one is the better option for fewer or more than that number of years?

A grant of $10,000 per year for x years adds up to $10{,}000x$ dollars. A starting balance of $5 doubled every year for x years can be modeled by the exponential expression $5 \cdot 2^x$. To find where these prize values are equal, set the expressions equal to one another: $10{,}000x = 5 \cdot 2^x$. To solve for x, we can define two functions as $f(x) = 10{,}000x$ and $g(x) = 5 \cdot 2^x$, graph both functions on the same coordinate grid, and find the point where the graphs intersect. At that point, the x-value is the same for both functions, and they both have the same function value ($f(x) = g(x)$). The x-value of the point of intersection is the x-value that makes $10{,}000x = 5 \cdot 2^x$ true, so it is the solution to the equation.

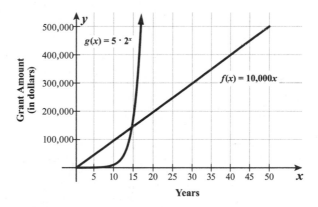

The two function graphs appear to intersect just to the left of 15, with a function value just under 150,000. The two prizes are worth close to the same amount after 15 years. For time periods less than 15 years, $f(x) = 10{,}000x$ has a greater value, so option (A) is the better choice. For time periods more than 15 years, $g(x) = 5 \cdot 2^x$ has a greater value, so option (B) is the better choice. This exponential function is continuing upward at a very fast rate here, so its endpoint, at $x = 20$, is not shown on this graph. The endpoint of $f(x) = 10{,}000x$ is shown, at $(50, 500{,}000)$.

Sylvie went on a Ferris wheel ride. At the top point of the wheel, she was 70 feet above the ground, and at the bottom point, she was 10 feet above the ground. Her ride lasted for a total of 3 minutes and 20 seconds, and made 5 complete revolutions. Write and graph a function modeling Sylvie's distance from the ground, $h(x)$, in feet, x seconds into her ride.

A Ferris wheel is in the shape of a circle, so the relationship of points along the circle (measured as a central angle measurement) to height above the ground is modeled by a sine or cosine function. Because the Ferris wheel rotates at a constant speed, time is directly proportional to the central angle measure, so the type of function that relates time to height above ground is also a sine or cosine function. Sylvie's ride begins at the minimum height, so let's use a cosine function, which starts at its maximum height when $x = 0$, but starts at its minimum height when reflected across its midline.

This problem is similar to Example 20 in Chapter 4, Lesson 4.6, which also involves translating a real-life situation to a trigonometric function equation and graph. The motion involved is still along the arc of a circle, but in this case, the motion is multiple revolutions of a complete circle.

The Ferris wheel completes 5 revolutions in 3 minutes and 20 seconds, or 200 seconds total, so it completes one full revolution in $200 \div 5 = 40$ seconds. The period of the cosine curve must be 40 seconds. The period of a cosine function of the form $y = a \cos b(x - c) + d$ is $2\pi/b$. We can set $2\pi/b$ equal to 40 to solve for b.

$2\pi/b = 40$
$2\pi = 40b$
$2\pi/40 = b$
$b = \pi/20$

The Ferris wheel carries Sylvie from a minimum height of 10 feet to a maximum height of 70 feet, so the minimum of the cosine curve is 10 and the maximum is 70. The distance between them is 60, so the amplitude of the cosine curve is 30. The midline is halfway between 10 and 70: $\dfrac{10 + 70}{2} = 40$.

A cosine function $y = a \cos b(x - c) + d$ with an amplitude of 30 ($|a| = 30$), a period of 40 ($b = \pi/20$), no horizontal shift ($c = 0$), a midline of 40 ($d = 40$), and reflection across the midline (negative a-value) is $y = -30 \cos \pi/20 \, x + 40$. The domain of our function is the time period that Sylvie was riding the Ferris wheel, from $x = 0$ to $x = 240$ seconds. The function modeling Sylvie's distance from the ground after x seconds on the Ferris wheel is $h(x) = -30 \cos \pi/20 \, x + 40$, for $0 \le x \le 240$. This function is graphed below.

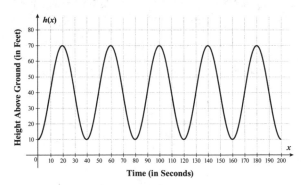

EXAMPLE 16 🔒

Zach wants to start selling cupcakes at his bakery, which currently sells only cakes and cookies. From his research, he has learned that the demand curve for cupcakes in the neighborhood is given by the function $n = -20p^2 + 500$, where n represents the number of cupcakes sold per day and p represents the price per cupcake. Zach's cost in ingredients and labor is $1 per cupcake. Write and graph a function $f(p)$ modeling Zach's daily profit in terms of p, the price per cupcake, in dollars. According to this function, at what price should Zach sell his cupcakes, to the nearest 5 cents? If it turns out that Zach only has the baking capacity for a maximum of 240 cupcakes per day, how does this change your answer for the price at which he should sell his cupcakes? Draw a new graph of the profit function, as a piecewise-defined function, to illustrate your answer.

The demand curve $n = -20p^2 + 500$ tells us how many cupcakes, n, will be sold at price p. The total revenue is the price per cupcake multiplied by the number of cupcakes sold, so revenue $r = p \cdot n = p(-20p^2 + 500)$. This simplifies as $r = -20p^3 + 500p$. Zach's cost for making n cupcakes at $1 per cupcake is n dollars. Substitute the expression for n in terms of p: cost $c = -20p^2 + 500$. Zach's profits are his revenue minus his cost, so $f(p) = r - c = (-20p^3 + 500p) - (-20p^2 + 500)$. Simplified, $f(p) = -20p^3 + 20p^2 + 500p - 500$. The graph of this function is shown below, with a domain of $0 \le p \le 5$.

The price per cupcake cannot be negative, so p must be greater than or equal to 0. When $p = 5$, the profit is equal to 0, which means that 0 cupcakes were sold. At that price, there is no longer any demand for them. So, the circumstances require a domain restriction to $0 \le p \le 5$.

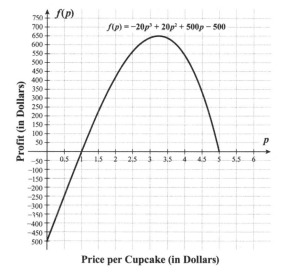

Price per Cupcake (in Dollars)

The maximum of this curve is about 650 and occurs around $p = 3.24$. To maximize his profit, Zach should set the price of a cupcake at $3.25.

Up until this point, this problem has been similar to Example 21 in Chapter 1, Lesson 1.6, where we also used a demand curve and cost information to write and graph a profit curve, solving for the price point for maximum profit. However, in this situation, there is a limiting factor, which changes the profit function to a piecewise-defined function.

If Zach's bakery can only bake a maximum of 240 cupcakes per day, then the profit function changes. The demand curve only reflects the number of cupcakes sold when its value is less than or equal to 240. Set $-20p^2 + 500$ less than or equal to 240 to represent this limitation.

$-20p^2 + 500 \leq 240$

$-20p^2 \leq -260$ Subtract 500 from both sides.

$p^2 \geq 13$ Divide both sides by −20, including flipping the inequality sign.

The variable p represents the price per cupcake, so it must be a positive number. You only need to solve for the positive square root of 13 in this case.

$p \geq 3.606$ Evaluate the positive square root of 13.

So, the number of cupcakes sold is given by $n = -20p^2 + 500$ only when price p is greater than or equal to \$3.61. When the price is \$3.60 or less, the number of cupcakes sold is 240, the maximum the bakery can produce, even though the demand is greater. For 240 cupcakes sold, the revenue is $240p$ and the cost is $240(1)$, so the profit is $240p - 240$. So, Zach's cupcake profits, $f(p)$, are given by the piecewise-defined function shown below.

$$f(p) = \begin{cases} 240p - 240, & 0 \leq p \leq 3.60 \\ -20p^3 + 20p^2 + 500p - 500, & 3.61 \leq p \leq 5 \end{cases}$$

The graph of the piecewise-defined profit function is shown below.

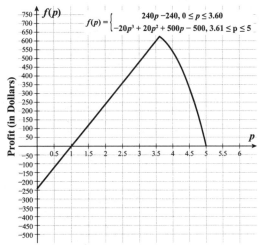

Price per Cupcake (in Dollars)

You can use your graphing calculator to track the function and/or calculate $f(3.65)$.

We can use the function to compare profits for a unit price of \$3.60 and a unit price of \$3.65.

$f(3.6) = 240(3.6) - 240 = 624$

$f(3.65) = -20(3.65^3) + 20(3.65^2) + 500(3.65) - 500 \approx 619$

In this case, with a maximum baking capacity of 240 cupcakes per day, Zach should set the price of a cupcake at \$3.60.

CHAPTER 7 PRACTICE QUESTIONS

Directions: Complete the following open-ended problems as specified by each question stem. For extra practice after answering each question, try using an alternative method to solve the problem or check your work.

1. The formula below represents the distance modulus of a star, or the difference between the star's apparent magnitude (how bright it appears from Earth), m, and its absolute magnitude (how bright it actually is), M, in terms of d, the distance of the star from Earth, in parsecs.

   $$m - M = 5\log_{10}(d - 5)$$

   Write a formula that gives d in terms of m and M. If one parsec is equal to about 3.26 light-years, then what equation produces distance D in light-years?

2. Since the year 2000, the number of regular subscribers to a certain magazine can best be described by the function $s(x) = 1200 \cdot 0.83^x$, where $s(x)$ is the number of subscribers x years after the year 2000. Is this function always increasing or always decreasing? What is the y-intercept of this function's graph, and what does it represent? If the number of magazine subscribers continues to follow this function, how many subscribers will there be in the year 2050?

3. Dianna has found that pants in the same labeled size made by different clothing brands have different sized waists, so she measures the waistband for each pair of pants she considers buying. Her ideal pants waistband measures 72 cm, but she still likes the fit of waistbands within 2.5 cm of this measure. Write an inequality that can be solved for w, the waistband measures Dianna prefers for her pants. If everyone preferred pants waistbands within 2.5 cm of their individual ideal waistband measure, m, then what inequality relates w and m?

4. Brian's yard is a rectangle, 40 feet long and 20 feet wide. He plans to put a walkway of width w feet around the border, within his yard, based on the area, a, of the middle portion of the yard enclosed by this walkway. Write and graph a function for w, the width of the walkway, in feet, in terms of a, the area of the enclosed yard space, in square feet.

5. Shervin flew a helicopter in a flight that rose above and dipped below an altitude of 5000 feet according to the function $h(t) = -t^4 + 10t^3 - 29t^2 + 20t$, where t represents time in minutes (since he first reached 5000 feet) and $h(t)$ represents the distance in feet above an altitude of 5000 feet. At what time was Shervin's helicopter exactly at an altitude of 5000 feet?

DRILL

6. The AC power outlet at Zawadi's home has a voltage $v(t)$ that follows the sine curve $v(t) = 120\sqrt{2} \cdot \sin(120\pi t)$, shown below, where t represents time in seconds.

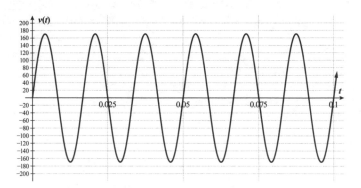

(a) The frequency of alternating current is defined in hertz, which are cycles per second. What is the hertz measurement for Zawadi's power supply?

(b) The peak-to-peak value of an AC voltage is twice the amplitude of the voltage sine curve. What is the peak-to-peak voltage of Zawadi's power supply, to the nearest tenth of a volt?

7. When the quarterback of the school football team caught the football, he saw a wide receiver 12 yards away from him. He watched the receiver run away from him for 1 second then threw the football at a speed 3.5 times the speed of the receiver, who continued to run at a steady rate. The receiver caught the ball 28 yards away from the quarterback, who had remained stationary. Write a function that relates t, the time in seconds since the quarterback caught the football, to r, the wide receiver's speed, in yards per second, for both the receiver's run and the ball's path through the air. Graph both functions on the same coordinate grid. What do the r- and t-values of the point of intersection represent in this situation? At what speed did the football travel through the air?

SOLUTIONS TO CHAPTER 7 PRACTICE QUESTIONS

1. $D = 3.26 \cdot 10^{\frac{m-M}{5}} + 16.3$

   To find an equation that shows the variable d equal to an expression in terms of m and M, you must solve the given formula for d. Use inverse operations to "undo" everything that has been done to d.

$\dfrac{m-M}{5} = \log_{10}(d-5)$	Divide both sides by 5.
$10^{\frac{m-M}{5}} = d - 5$	Use the definition of a logarithm to rewrite as an exponential equation.
$10^{\frac{m-M}{5}} + 5 = d$	Add 5 to both sides.

   The formula that gives distance d, in parsecs, in terms of m and M is $d = 10^{\frac{m-M}{5}} + 5$.

   If one parsec is equal to 3.26 light-years, then d parsecs are equal to $3.26d$ light-years.

   Distance, D, in light-years, is equal to $3.26d$. Set D equal to 3.26 times the expression we found for d. The formula that gives distance D, in light-years, in terms of m and M is $D = 3.26(10^{\frac{m-M}{5}} + 5)$, or $D = 3.26 \cdot 10^{\frac{m-M}{5}} + 16.3$.

2. **always decreasing, y-intercept represents 1200 subscribers in 2000, and there will be no subscribers in 2050.**

   This is an exponential function with a base that is less than 1 ($0.83 < 1.00$) and a positive coefficient, so it is always decreasing. For each increase of x by 1, the function value is multiplied by 0.83, which produces a smaller function value. The magazine has been losing subscribers since the year 2000.

   The y-intercept (or $s(x)$-intercept) occurs when $x = 0$, so substitute 0 for x and solve for $s(x)$.

   $s(0) = 1200 \cdot 0.83^0 = 1200 \cdot 1 = 1200$

   The y-intercept for this function is 1200. The variable x is years after 2000, and the function value, $s(x)$, is the number of subscribers, so the y-intercept tells you that there were 1200 subscribers to this magazine in the year 2000.

   The year 2050 is 50 years after 2000, so use $x = 50$. When $x = 50$, $s(x) = 1200 \cdot 0.83^{50} \approx 0.1$. The $s(x)$-value represents the number of subscribers, so we must round to the nearest whole number, which is 0. If the function pattern continues, the magazine will have no subscribers by 2050.

3. **$|w - m| \leq 2.5$**

Dianna prefers pants waistbands that are within 2.5 cm of 72 cm, whether greater than or less than 72 cm. Use an absolute value to represent the difference between the actual waistband measure and 72 cm, as a positive value: $|w - 72|$. Set this positive difference less than or equal to 2.5, to represent Dianna's preference that the difference be within 2.5 cm.

$|w - 72| \leq 2.5$

To solve for w, we could graph $f(w) = |w - 72|$ and find the w-values for $f(w)$-values less than or equal to 2.5. Or, we could solve algebraically for cases where $(w - 72)$ is positive and negative.

If $w - 72 \geq 0$:   If $w - 72 < 0$:
      $w - 72 \leq 2.5$ $-(w - 72) \leq 2.5$
      $w \leq 74.5$ $-w + 72 \leq 2.5$
                                                  $-w \leq -69.5$
                                                   $w \geq 69.5$

Dianna's preferred pants waistband measurements are given by $69.5 \leq w \leq 74.5$.

If everyone preferred pants with waistbands within 2.5 cm of m, a personal ideal waistband measure, then the absolute difference between w and m must be less than or equal to 2.5 cm: $|w - m| \leq 2.5$.

4.   **$w = \sqrt{\dfrac{a}{4} + 25} + 15$; see graph on following page**

To better visualize the problem, draw a sketch of the yard with walkway. Brian's entire yard is 40 feet long and 20 feet wide. The walkway is within the yard, at its border, with a width everywhere (on every side) of w feet.

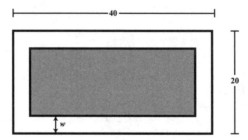

The area of the middle portion of the yard enclosed by the walkway (the shaded part in the sketch), in square feet, is defined as a. To calculate this area, you would multiply the length and width of this central rectangle. Use the dimensions given in the diagram to write expressions for the length and width. The length of this central rectangle is the full 40-foot length of the yard minus the walkway space on either side, or $40 - 2w$. Similarly, the width of this enclosed portion of the yard is $20 - 2w$. Write an equation using the fact that a rectangle's area is equal to its width times its length.

$a = (20 - 2w)(40 - 2w)$
$a = 800 - 120w + 4w^2$ Expand using FOIL.

We want an equation that gives *w* in terms of *a*, so solve this equation for *w*.

$4w^2 - 120w + 800 = a$	Rearrange terms.
$4(w^2 - 30w + 200) = a$	Factor out 4 on the left side of the equation.
$w^2 - 30w + 200 = a/4$	Divide both sides by 4.

To solve for *w* alone, we need to write a perfect square of a binomial containing *w*. To do that, we need to complete the square for $w^2 - 30w$. Half of $-30w$ is $-15w$, so we want to build $(w - 15)^2$, or $w^2 - 30w + 225$.

$w^2 - 30w = a/4 - 200$	Subtract 200 from both sides.
$w^2 - 30w + 225 = a/4 - 200 + 225$	Add 225 to both sides, to complete the square on the left.
$(w - 15)^2 = a/4 + 25$	Rewrite the quadratic as the square of a binomial.
$w - 15 = \pm\sqrt{\dfrac{a}{4} + 25}$	Take the square root of both sides.
$w = \pm\sqrt{\dfrac{a}{4} + 25} + 15$	Add 15 to both sides.

Although this is a relation, it is not a function, because of the ± symbol. The graph of this relation is a sideways parabola, but the situation describes a function relationship, so you must decide which portion of the parabola represents the situation. As the area of the enclosed portion in the middle of the yard increases, the width of the walkway decreases, so the function should be always decreasing. So, the lower half of the parabola, or $w = -\sqrt{\dfrac{a}{4} + 25} + 15$ matches the situation.

However, you must also consider what domain is appropriate to the situation. The area of the central portion of the yard must be between 0 and the full area of the yard, $40 \cdot 20 = 800$ square feet, so $0 \le a \le 800$. (Realistically, the domain might be further reduced, to provide at least some walkway or at least some central area within the walkway, but we do not have any further information to specifically define those domain restrictions.) The graph of $w = -\sqrt{\dfrac{a}{4} + 25} + 15$ with domain $0 \le a \le 800$ is shown below.

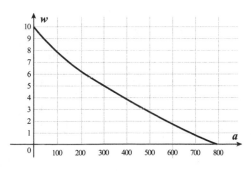

5. **0, 1, 4, and 5 minutes**

In this situation, $h(t)$ is the helicopter's height above an altitude of 5000 feet, so negative values of $h(t)$ represent locations of the helicopter below the 5000-foot level. The helicopter is at exactly 5000 feet in altitude whenever $h(t) = 0$. This function is a polynomial function, so when it is fully factored, it clearly indicates its zeros.

First, pull out the common factor of $-t$.

$-t^4 + 10t^3 - 29t^2 + 20t = -t(t^3 - 10t^2 + 29t - 20)$

According to the Rational Root Theorem, zeros of the cubic $t^3 - 10t^2 + 29t - 20$ will be in the set $\pm\{1, 2, 4, 5, 10, 20\}$. Because the variable t in this situation is always positive (time passed), the zeros must be positive. Let's test $t = 1$ in the original function, using the Remainder Theorem.

$h(1) = -1^4 + 10(1^3) - 29(1^2) + 20(1)$
$\quad\quad = -1 + 10 - 29 + 20$
$\quad\quad = 0$

Because $h(1) = 0$, $(t - 1)$ is a factor of $h(t)$. Now we must factor $(t - 1)$ out of $(t^3 - 10t^2 + 29t - 20)$, using long division.

$$
\begin{array}{r}
t^2 - 9t + 20 \\
t-1 \overline{)\, t^3 - 10t^2 + 29t - 20} \\
\underline{-(t^3 - t^2)} \\
-9t^2 + 29t \\
\underline{-(-9t^2 + 9t)} \\
20t - 20 \\
\underline{-(20t - 20)} \\
0
\end{array}
$$

So, $h(t) = -t(t - 1)(t^2 - 9t + 20)$, which fully factors as $h(t) = -t(t - 1)(t - 4)(t - 5)$. In this form, we can see that the zeros of the function $h(t)$ are 0, 1, 4, and 5. The helicopter is at an altitude of 5000 feet 0, 1, 4, and 5 minutes after initially reaching that level.

6. **339.4 volts**

In the sine graph shown, 6 full cycles are completed in 0.1 second. We can set up and solve a proportion to find the number of cycles completed in 1 second.

$6/0.1 = x/1$	Write the equivalent ratios of cycles to seconds.
$6 = 0.1x$	Cross-multiply.
$60 = x$	Multiply both sides by 10.

There are 60 cycles per second, so Zawadi's power supply has a frequency of 60 hertz.

The sine curve appears to reach a maximum of 170 and a minimum of −170, so the graph indicates an amplitude of about 170, or a peak-to-peak of 340 volts. However, we can find a more precise answer using the given equation. The amplitude of a sine function of the form $y = a \sin b(x - c) + d$ is $|a|$. In the function $v(t) = 120\sqrt{2} \cdot \sin(120\pi t)$, $a = 120\sqrt{2}$, so $|a| = 120\sqrt{2}$. We must multiply this amplitude by 2 to get the peak-to-peak voltage: $240\sqrt{2}$. Use a calculator to find that $240\sqrt{2} = 339.4$, to the nearest tenth. The peak-to-peak value of Zawadi's power supply is 339.4 volts.

7. **see functions and graph below; football travels at 28 yards per second**

The problem concerns the relationship between distance, rate, and time, which is given by $d = rt$. In this case, you must write a function for time in terms of rate and distance, so rewrite this formula as $t = d/r$.

The wide receiver ran at a steady rate, from when the quarterback caught the football ($t = 0$, because t is defined as seconds since the quarterback caught the ball) until he caught the pass. His running speed, in yards per second, is defined as r. To find the distance he ran, we can subtract his original distance from the quarterback, 12 yards, from his final distance of 28 yards from the quarterback, when he caught the ball.

$d = 28 - 12 = 16$

So, the function $t = 16/r$ describes the receiver's run.

The football travels a total distance of 28 yards through the air, at a speed 3.5 times the speed of the wide receiver, or $3.5r$. So, the time the football is in the air is equal to $\dfrac{28}{3.5r}$, which can be simplified as $8/r$. However, t is the time since the quarterback caught the football, and he waited 1 second before throwing the ball. So, for the football, $t = 1 + 8/r$.

Both functions are graphed on the coordinate grid below.

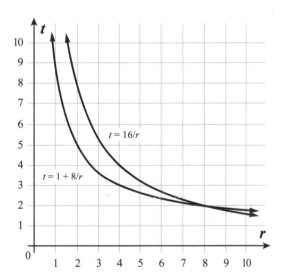

The graphs intersect at the point (8, 2), where $r = 8$ and $t = 2$. Both functions were written to express the situation where the ball and the wide receiver each traveled the necessary distance to end up at the same location, for a successful pass. But, in order for the pass to succeed, they must also arrive there at the same time. The r-value that produces the same t-value in both functions is 8. So, the wide receiver's speed was 8 yards per second. The t-value of 2 means that the receiver caught the ball 2 seconds after the quarterback caught the ball.

The quarterback threw the ball with a speed of 3.5 times the running speed of the wide receiver, so the football traveled through the air at a speed of $3.5 \cdot 8 = 28$ yards per second.

REFLECT

**Congratulations on completing Chapter 7!
Here's what we just covered.
Rate your confidence in your ability to**

- Manipulate given equations to solve for certain variables or to reveal properties of functions

  ① ② ③ ④ ⑤

- Determine key features of functions from equations, tables, and graphs, and interpret those features in the context of the situations described by the functions

  ① ② ③ ④ ⑤

- Model described situations with equations and inequalities, and use them to solve problems

  ① ② ③ ④ ⑤

- Model described situations with graphs, and use them to solve problems

  ① ② ③ ④ ⑤

If you rated any of these topics lower than you'd like, consider reviewing the corresponding lesson before moving on, especially if you found yourself unable to correctly answer one of the related end-of-chapter questions.

 Access your online student tools for a handy, printable list of Key Points for this chapter. These can be helpful for retaining what you've learned as you continue to explore these topics.

Chapter 8
Inferences and Conclusions from Data

8

GOALS By the end of this chapter,
you will be able to:

- Understand the difference between sample surveys, experiments, and observational studies, and identify methods that provide representative sample populations for each of these

- Find the standard deviation or estimated standard deviation, as appropriate, for a data set, and understand the relationship between standard deviations, mean, and the normal distribution curve

- Use proportional reasoning, given ratios in a large representative sample population, to make inferences about the target population

- Understand the difference between discrete uniform, continuous uniform, normal, skewed, and bimodal probability distributions

- Use area under a continuous probability distribution curve, either uniform or normal, to calculate probabilities, when appropriate to the situation

- Use simulations to assess how well experimental results match a given model and to determine when results are statistically significant

- Calculate the confidence interval and margin of error for a given sample proportion at various confidence levels and for various sample sizes

- Use statistics and probability concepts to interpret and evaluate given statements and to analyze options in real-world situations

Lesson 8.1
Data Collection

Statistics is the collection, analysis, and interpretation of numerical data, typically for the purpose of making inferences and conclusions about a larger population through use of a sample subset of that population. It would be very time-consuming, if not impossible, to count every individual soccer player in the United States. However, using statistics, you can estimate the total number of Americans who play soccer based on a smaller sample of the general population.

Probability is an important concept behind statistical analysis, because the true parameters of the larger population (the **target population**) are typically unknown. We cannot make absolute statements about those parameters, but we can make conclusions in terms of how likely they are. (A **parameter** is some numerical characteristic of the population, such as its mean or range.)

The selection of the sample population for a study is very important. To obtain the most accurate results in a study of some **treatment**, or factor that we want to study in terms of its effect, we must attempt to equalize other factors as much as possible. For a fair test of the treatment, we must choose a representative sample of the target population, to minimize introduction of other factors that might influence results. The only way to extract a representative sample is through a random sampling method, in which each member of the target population has as equal a chance of being chosen as possible.

In Heather's family, one person is chosen each night to take out the trash. What are some examples of fair selection processes, in which each member of the family has an equal probability of being chosen?

Each member of the family could write his or her name on a piece of paper and put the paper into a hat. Someone could then draw a name from the hat, without looking.

Each person could draw a straw from a group of straws (as many straws as there are members of the family), one of which is shorter than the others (but with the bases hidden from view so that no one can tell which is shorter). The person with the short straw takes out the trash.

Each person could write down a number between 1 and 100, and the person whose number first appears in a list of randomly generated numbers between 1 and 100 has to take out the trash. One way to randomly choose a number between 1 and 100, or a series of such numbers, is by using a random number generator (generally a computer program).

An option of choosing the person whose number is closest to a single randomly generated number would not be fair, because certain numbers, such as 1, 2, 99, and 100 are generally less likely to be closest to the randomly generated number.

Such methods work for small populations, in everyday choices for fairness. For very large populations, you must use other methods, but the goal is still to make probabilities of being chosen as similar as possible across the entire population, for a random sample. If a sample is not chosen randomly, the method of choosing members of the sample introduces bias. **Bias** is the tendency of a measurement process to favor one group or outcome over others, producing a lopsided view of the population as a whole.

Three ways of obtaining a sample are sample surveys, experiments, and observational studies.

SAMPLE SURVEYS

A **sample survey** is a survey of members of a smaller sample group from the target population, with the goal of learning something about the target population. For example, you might poll randomly collected samples of Americans to ask if they play soccer, in the hopes of extrapolating to find the total number of soccer players in the country.

Bias in the sample selection can also be viewed as a method of exclusion. Surveying only those Americans who have college degrees excludes all of the Americans who do not have college degrees. This means the sample is not representative of the entire target population.

The sampling method should be random, so that the sample subset is representative of the target population. If a survey to determine American opinions on the importance of education is limited to a sample group of only people who have college degrees, then the results of the survey will be biased; they will not reflect the opinions of all Americans. Even when the specific correlation is not obvious, a certain subset of the population, sharing some characteristic, may also tend to share other characteristics, affecting how they will respond to the given survey.

EXAMPLE 2

Principal Ruiz wants to use student opinion to choose one of two options to become the school's new mascot. Which of the following sample methods will give him the most accurate reflection of overall student opinion on the issue?

A) Poll the 80 students that belong to sports teams at the school.

B) Poll the 150 students in the freshman class.

C) Poll the first 30 students who arrive at school the next morning.

D) Poll every fifth student from an alphabetized list of all 640 students at the school.

A group of just the sports team members or just the freshmen does not accurately represent the entire student body. Students within one of those groups may vote differently, on the whole, from other students in the school, for one reason or another.

Even though the 150 students in the freshman class constitute a larger sample population, they are not representative of the entire student body and may skew the data. The 128 students chosen randomly are a better sample population for the survey, especially considering that the sample size is not substantially different.

The first 30 students who arrive at school the next morning provide more of a random sample, as do the group of students made up of every fifth student on an alphabetized list of all students. To compare these two options, look at the number of people being surveyed. Every fifth student from a list of 640 students means 1/5 of 640, or 128. This is a much larger sample size than 30, so it is much more likely to provide an accurate representation of the opinions of the full student body. Also, the students arriving earliest at the school may share certain characteristics and provide less of a representative sample. Principal Ruiz should poll every fifth student from an alphabetized list of all students at the school.

The most important factors determining the accuracy of a sample survey, in terms of sampling methods, are the randomization of the selection process and the sample size. A sample should ideally be both randomly selected and large.

There are other ways in which a survey or its sample selection may be biased. If a survey is open to anyone who wants to take it, then there is a **voluntary bias**, because the types of people who choose to take the survey may share similar characteristics and not represent the opinions of the entire population. The ways in which survey questions are worded can also skew the data, by encouraging certain answers.

EXAMPLE 3

Anna wants to determine how town residents feel about a recent proposal to replace Maplewood Park with a mall. In her survey, she included the following question: "Maplewood Park has been a valued part of our town for the past 175 years. Would you rather preserve this public space or allow the proposed Maplewood Mall to take it over?" How is this question biased? How might Anna reword the question so that it is unbiased?

This question shows the park in a positive light, as both a "valued part of our town for the past 175 years" and as a "public space" to be "preserve(d)." In this way, Anna is encouraging respondents to choose the preservation of the park over the building of the mall.

An unbiased version of the question should present both options on an equal level, without endorsing, directly or indirectly, either option. Anna might simply rewrite the question as "Would you prefer to keep Maplewood Park or replace it with Maplewood Mall?"

Here is how you may see sampling errors on the SAT.

Freewheeling Four is a company that designed a new wheelchair. They want to determine whether this wheelchair will be popular among senior citizens, so they decided to do product testing at five different community centers around the country. They invited all interested community members to attend a demonstration of the wheelchair in the backyard of each community center. Through surveys, Freewheeling Four found that 58% of the senior citizens who attended the demonstrations were interested in purchasing the wheelchair. Based on this finding, would Freewheeling Four be justified in claiming that the majority of senior citizens would likely purchase the new wheelchair?

A) Yes; Freewheeling Four's study used a representative sample of the target population, indicating that the conclusion is accurate.

B) Yes; the majority of senior citizens who attended the demonstrations indicated that they would likely purchase the new wheelchair.

C) No; the senior citizens were just one part of the group of all community members who attended the demonstrations, so they do not accurately represent the target population.

D) No; while 58% of the senior citizens who attended the demonstrations indicated that they would likely purchase the new wheelchair, the sample failed to accurately represent the target population.

EXPERIMENTS

Another form of data collection is through experiments. If Michael tosses a coin repeatedly, recording each result as heads or tails, this is an experiment that can be used to determine the probability that a coin tossed in the air will land heads up. The sample size, in this case, is the number of times Michael tosses the coin. This kind of experiment is by its nature random, because the person running the experiment has no way of affecting the outcomes of the tosses. Because this experiment is already random, the sample size becomes very important for determining the accuracy of the results.

EXAMPLE 4

In his first experiment, Michael tossed a penny 5 times, with the result {T, T, T, H, T}, where H represents heads and T represents tails. In his second experiment, he tossed a penny 20 times, with the results {H, T, T, T, T, H, H, T, H, T, H, H, H, T, T, H, H, H, H, H}. What was the experimental probability of tails in the first experiment? What was the experimental probability of tails in the second experiment? If Michael combines all the data from both series of coin tosses, as one complete data set, what is the probability of tails in the complete experiment?

In the first experiment, the penny landed on tails 4 out of the 5 times it was tossed, so the probability of tails was 4/5, or 0.8. In the second experiment, the penny landed on tails 8 out of the 20 times it was tossed, so the probability of tails was 8/20, or 0.4. If we combine all the coin toss data, the penny landed on tails a total of 4 + 8 = 12 times, out of a total of 5 + 20 = 25 tosses. For Michael's complete experiment, the probability of the penny landing on tails was 12/25, or 0.48.

Notice that the experimental results get closer to the theoretical probability of a penny landing on tails, 1/2 or 0.5, as the sample size increases. For a very small sample size, it is actually unlikely that a coin will land on tails exactly half the time. On the other hand, for a very large sample size, the portion of the time that the penny lands on tails will be very close to 0.5.

> The average of the results of performing the same experiment a large number of times will be close to the expected value, or theoretical probability. And, it will continue, overall, to become closer to that expected value as more trials are performed.

It's also very unlikely that a coin will land on tails exactly half the time in a very large number of tosses. However, the ratio of outcomes of tails to total number of outcomes becomes very close to 0.5.

Experiments are not always related to processes with known (theoretical) probabilities. When a researcher tests a new medication on a sample group of people, the method is again called an experiment. As with sample surveys, it is important to randomly select the sample group from the target population. Also, for a controlled experiment, such as testing a new medication, there must also be a **control group**, which does not receive the treatment being tested, so that the researcher can compare results. The effects of the treatment are only measurable in reference to a baseline of how the subjects would fare without the treatment.

AT A

Freewheeling Four visited only five locations nationwide, which is a very small sample size for the entire country. The locations visited were community centers, but not all senior citizens visit community centers. (It's possible, for example, that senior citizens with more money are more likely to purchase the new wheelchair, while being less likely to visit community centers.) Finally, by having their demonstrations in the backyard, with only those curious to watch it attending, they introduced a voluntary bias.

So, even though the majority of senior citizens who attended the demonstrations would be interested in purchasing the new wheelchair, the sample does not accurately represent the target population of all senior citizens in the United States. The correct answer is (D).

EXAMPLE **5**

Neal tested his vitamin formula on a randomly selected group of 50 subjects who had colds. The cold symptoms went away within a week for 34 of the members of this sample population. Neal advertises his formula by saying, "68% of people with colds who take my formula recover within a week." If a group of 60 different subjects with colds, also randomly selected from the same population but not given Neal's formula, included 39 people whose symptoms went away within a week, how might this affect someone's view of Neal's results?

Neal's statement may be true, because 68% of his sample population recovered within a week after taking his vitamin formula. However, in the second group, 39/50 = 65% of subjects with colds who did not take Neal's formula also recovered within a week. These percentages are not very far apart, so there doesn't appear to be much of an impact of the vitamin formula. Without a point of reference for comparison, someone might be impressed by Neal's statement. With the control results, he or she would likely find little if any correlation between consumption of the vitamin formula and recovery rate.

Notice that in the case of Neal's experiment, the target population is people with colds, so any sample, whether for the test group or for the control group, must come from the population of people with colds. It would not make sense to give healthy test subjects the vitamin formula, nor to include already healthy people in the control group. In a study of how a formula affects cold recovery rates, the only meaningful comparison is of people with colds who take the formula and people with colds who do not take the formula, with all other characteristics or factors equalized as much as possible between the two groups (achieved largely through random sample selection).

OBSERVATIONAL STUDIES

An **observational study** is similar to an experiment in that the researcher is studying the effect of some variable or treatment on some other variable, but is different in that it only involves observations. While an experiment is a method of applying a treatment (or treatments) to one or more groups and comparing the results with a control group, an observational study is a method of collecting and interpreting observations, without interfering with the subjects or variables in any way. The researcher is still responsible for randomly selecting a representative sample population from the target population, but the separation of subjects into a treatment group and a control group is beyond his or her control.

EXAMPLE 6

Lucille hypothesizes that a certain pesticide has negative impacts on the health of those who come into regular contact with it. To test this theory, she uses an observational study with a sample group of fruit pickers who regularly handle fruit at farms using this pesticide, as well as a control group of people who do not come in direct contact with the pesticide. What are the advantages of using an observational study instead of a controlled experiment in this case? How might using this observational study create bias that skews the results?

In a controlled experiment, Lucille would "treat" certain randomly selected subjects with the pesticide. Considering that there are potential health risks associated with exposure to the pesticide, it is unethical to expose people to the pesticide who would not otherwise be exposed. By using an observational study, Lucille can compare the health of people who are in regular contact with the pesticide to the health of those who are not, without subjecting anyone who is not normally in contact with it to the pesticide.

Lucille has no control over the composition of the population of fruit pickers at these farms, so she cannot ensure that they are a representative sample of the target population, or that they are similar in composition to the randomly selected control group. The fruit pickers at the farms using the pesticide may share certain characteristics, such as ethnicity or economic level, that can make it harder to draw a direct correlation between health and pesticide exposure, because there may be other variables or factors influencing the results.

Lesson 8.2
Means and Measures of Variability

Statistics is used not just to measure the mean of a set (or other measures of central tendency), but also the variability of the data in that set, for example, in terms of how far it is from the mean. Two data sets may have the same mean but different variability, even if containing the same number of elements. The sets {9, 10, 10, 10, 10, 11} and {2, 5, 6, 9, 18, 20} each contain six elements and each have a mean of 10, but the data in the second set varies more from the mean.

Because the lower quartile marks the 25% point in the data and the upper quartile marks the 75% point, the interquartile range encompasses approximately half of the total elements in the set. For very small data sets, the interquartile range is not likely to contain exactly half of the elements in the set, but for large data sets, it contains very close to, or exactly, 50% of the elements in the set.

There are various **measures of variability** that give you some sense of the dispersion, or spread, of the data. The simplest of these is the range. The set {9, 10, 10, 10, 10, 11} has a range of 2, and the set {2, 5, 6, 9, 18, 20} has a range of 18. Another measure of variability is the interquartile range. The interquartile range of {9, 10, 10, 10, 10, 11} is 0, and the interquartile range of {2, 5, 6, 9, 18, 20} is 13. By only measuring the range of the middle half of the data, the interquartile range provides a better sense of the behavior of data in the set, as compared to the range, which can be greatly affected by very small or large values that are unusual within the set.

A more precise measure of variability is variance, because it compares each element in the set to the arithmetic mean of the set, which is represented by the Greek letter μ (mu). The **variance**, σ^2, is the mean of the squared deviations of all data points from the mean of the data set. In other words, to find the variance, we find the difference between each data point in the set and the mean, square each of those differences, and then find the mean of the squares.

The formula for variance is:

$$\sigma^2 = \frac{\sum_{i=1}^{n}(x_i - \mu)^2}{n},$$

where x_i represents the ith element of the data set, n represents the number of elements in the set, and μ represents the mean of the set.

The purpose of squaring the differences is to eliminate the variation between positive and negative values, and focus instead on the distance of each element from the mean of the set.

For the set {9, 10, 10, 10, 10, 11}, the variance is given by

$$\frac{(9-10)^2 + (10-10)^2 + (10-10)^2 + (10-10)^2 + (10-10)^2 + (11-10)^2}{6},$$ which

simplifies to 1/3, or $0.\overline{3}$. Variance is always expressed in square units for data points in

units, because it is an average of the squares of deviations.

For the set {2, 5, 6, 9, 18, 20}, the variance is given by

$$\frac{(2-10)^2 + (5-10)^2 + (6-10)^2 + (9-10)^2 + (18-10)^2 + (20-10)^2}{6},$$ which is

equal to 45 square units. The variance for this set is much greater than the variance for

the first set.

Often, a statistician is analyzing data for a sample from a target population, not data for the entire target population. In this case, we should use the formula for **estimated variance**, s^2. The estimated variance is, confusingly, sometimes called **sample variance**, but it is an estimate of the variance of the entire population based on data from a sample.

The formula for estimated variance is:

$$s^2 = \frac{\sum_{i=1}^{n}\left(x_i - \bar{x}\right)^2}{n-1},$$

where x_i represents the ith element of the sample data set,
n represents the number of elements in the sample set,
and $\bar{x}$ represents the mean of the sample data set.

Notice that, in the estimated variance formula, the sum of squares is divided by $(n-1)$ instead of n. The smaller denominator produces a larger variance value, to account for the possibility that data for the target population varies more than what is shown within the sample set.

If the set {2, 5, 6, 9, 18, 20} is a sample from a larger population that we are studying, then we would estimate the variance of the population using the estimated variance formula with data from this sample set.

$$s^2 = \frac{(2-10)^2 + (5-10)^2 + (6-10)^2 + (9-10)^2 + (18-10)^2 + (20-10)^2}{6-1}$$

Simplified, the estimated variance for the population, based on the given sample set, is 54 square units.

The Greek letter σ that we use to represent standard deviation is a lower-case sigma. The upper-case sigma, Σ, represents a sum, as used in the standard deviation formula.

If we take the square root of the variance of a data set, the result is the **standard deviation** of the data set, indicated by the Greek letter σ (sigma). Likewise, the square root of the estimated variance is the **estimated standard deviation**, also sometimes called the sample standard deviation. The estimated standard deviation is the estimate of the standard deviation for the entire population based on data from a sample.

Standard Deviation:

$$\sigma = \sqrt{\frac{\sum_{i=1}^{n}\left(x_i - \mu\right)^2}{n}}$$

Estimated Standard Deviation:

$$s = \sqrt{\frac{\sum_{i=1}^{n}\left(x_i - \bar{x}\right)^2}{n-1}}$$

The heights of the members of the high school girls' volleyball team are {5.2, 5.4, 5.4, 5.5, 5.5, 5.5, 5.7, 5.8, 6.0, 6.0}, in feet. The heights, in feet, of the members of the high school boys' basketball team are {5.6, 5.9, 5.9, 6.0, 6.0, 6.1, 6.1, 6.3, 6.4, 6.6, 6.7, 6.8}. Compare the means and standard deviations for the two teams. What do these indicate about the heights of players on each of these two teams?

To calculate the mean height of a player, add all heights for that team and divide the total by the number of players on the team. Notice that the number of players is different for the two teams: 10 girls on the volleyball team and 12 boys on the basketball team.

Girls' volleyball team:

$$\mu = \frac{5.2 + 5.4 + 5.4 + 5.5 + 5.5 + 5.5 + 5.7 + 5.8 + 6.0 + 6.0}{10} = 56/10 = 5.6$$

Boys' basketball team:

$$\mu = \frac{5.6 + 5.9 + 5.9 + 6.0 + 6.0 + 6.1 + 6.1 + 6.3 + 6.4 + 6.6 + 6.7 + 6.8}{12} = 74.4/12 = 6.2$$

For each team, the data given is for the entire population, so we are calculating the actual standard deviation, not an estimated standard deviation. Use the formula

$$\sigma = \sqrt{\frac{\sum_{i=1}^{n}\left(x_i - \mu\right)^2}{n}}.$$

Girls' volleyball team: $\sigma = \sqrt{\dfrac{\sum_{i=1}^{n}\left(x_i - 5.6\right)^2}{10}}$

Boys' basketball team: $\sigma = \sqrt{\dfrac{\sum_{i=1}^{n}\left(x_i - 6.2\right)^2}{12}}$

Now is a good time to use a calculator or computer program that will calculate the standard deviation for you. For the girls' volleyball team, the population variance is 0.064, and the population standard deviation is about 0.25. For the boys' basketball team, the population variance is 0.12167, and the population standard deviation is about 0.35.

Here is how you may see mean and standard deviation on the SAT.

Barbara's company had record profits last year. To thank her employees for their hard work, she divided a portion of the profits equally among all of her employees, so each employee received a bonus of $5,000. What effect does this bonus have on the mean and standard deviation of the employees' incomes for that year?

A) It would have no effect on the mean but would increase the standard deviation.

B) It would increase the mean but decrease the standard deviation.

C) It would increase the mean but have no effect on the standard deviation.

D) It would increase the mean and the standard deviation.

The mean height of a player on the boys' basketball team, 6.2 feet, is 0.6 foot greater than the mean height of a player on the girls' volleyball team, 5.6 feet. The heights of players on the boys' basketball team have a greater standard deviation, 0.35 foot, than those of players on the girls' volleyball team, 0.25 foot. In other words, the boys' basketball team members tended to be taller than the girls' volleyball team members, and there was more variation in their heights (they were more spread out from the mean).

Lesson 8.3
Frequency Distributions

One way to analyze data is to look at the relative frequencies of various outcomes, or data points. Let's compare the data distributions for a simple coin toss and a study of weights of grapes.

Suppose that an experiment consisting of 10,000 coin tosses has 4980 outcomes of heads and 5020 outcomes of tails. A frequency table for the coin toss experiment would look like this:

Outcome	Frequency
Heads	4980
Tails	5020

The frequencies could be shown as a bar graph, allowing you to visually compare the frequency with which each result occurred.

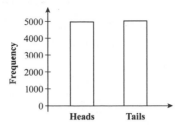

Suppose a researcher weighed 10,000 grapes of a certain variety and categorized the data as number of grapes per each half-gram weight interval, with the results shown in the following frequency table.

Even though we do not know how many employees are in Barbara's company or what their incomes are, or even the mean of their incomes, we can still determine how a constant added to every element in a population would affect the mean and standard deviation.

Each employee's income will increase by $5,000 for that year, so the mean of all their incomes will also increase. In a formula for the mean, the sum of all employee incomes will get an added $5000n$ ($5,000 times n employees in the company), and the total, when divided by n, yields an additional 5000 to what would have been the mean without the bonuses. The mean income will increase by exactly $5,000.

Each of the income data points in the set is increased by the same amount, so the deviation of data points from the mean (which is also increased by the same amount) is not affected. In the formula, each deviation that is squared is the difference $(x_i - \mu)$. Each x_i, or individual employee income, increased by 5000, but the mean, μ, also increased by 5000. The difference $((x_i + 5000) - (\mu + 5000))$ simplifies to $(x_i - \mu)$, the value of a deviation without a bonus. None of the deviations were affected, so the standard deviation remains the same.

The correct answer is (C).

Weight, x (in Grams)	Frequency (Number of Grapes)
$2.0 \le x < 2.5$	10
$2.5 \le x < 3.0$	251
$3.0 \le x < 3.5$	487
$3.5 \le x < 4.0$	552
$4.0 \le x < 4.5$	900
$4.5 \le x < 5.0$	1298
$5.0 \le x < 5.5$	1793
$5.5 \le x < 6.0$	1740
$6.0 \le x < 6.5$	1349
$6.5 \le x < 7.0$	845
$7.0 \le x < 7.5$	520
$7.5 \le x < 8.0$	234
$8.0 \le x < 8.5$	21

As with the coin toss results, the information in the frequency table for grape weights can be represented graphically as a frequency distribution. In this case, however, the outcomes are intervals in a continuous range of values, so the frequency distribution graph is a histogram.

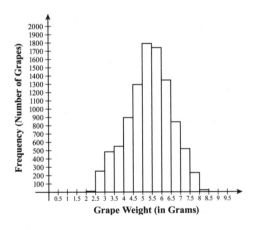

Unlike in the coin toss, where both outcomes occurred with about the same frequency, here it ranges from 10 grapes for an outcome (grape weight) of between 2.0 and 2.5 grams to 1793 grapes for an outcome of between 5.0 and 5.5 grams. Most of the grapes weighed fall within the middle range of weights, centered at about 5.0 to 5.5 grams. Because this is a large data set (and supposing it is representative of the grape variety), we know that if we pick a grape of this variety at random, it is likely to weigh about 5 or 6 grams and very unlikely to weigh 2 grams or 8 grams.

> If a sample is large and representative of the target population, then ratios of data in the sample set are likely to be close to the same as ratios of corresponding values in the target population. We can use this proportional reasoning to make inferences about the target population.

EXAMPLE 8

Angus surveyed a random sample of 90 students at his school, asking whether they preferred potato chips, tortilla chips, or popcorn. All 90 students responded, with 36 favoring potato chips, 21 favoring tortilla chips, and 33 favoring popcorn. If there are a total of 738 students at Angus's school, what is a good estimate of the number of students at the school who prefer popcorn to potato chips or tortilla chips?

Out of the 90 students Angus surveyed, 33 students preferred popcorn, so the ratio of popcorn fans to the total number surveyed is 33/90, or 11/30. The sample size is relatively large in relation to the target population size, and it was a randomly chosen sample, so it should be representative of the full student population. So, around the same ratio, 11/30, of the total number of students, 738, should prefer popcorn.

11/30 · 738 = 270.6

There cannot be a fraction of a person, so we must round off to a whole number, which is appropriate anyway for an estimate. It is likely that about 271 students at Angus's school prefer popcorn to potato or tortilla chips.

Keep in mind that this is just an estimate, and it's pretty much equally likely that 270 students at the school prefer popcorn. It's quite possible that the number is instead 282. There's even some slight possibility that Angus happened to include the only 33 popcorn fans in the school in his survey, but that is highly unlikely. Rather than just finding a general estimate for a value (such as number of students at a school who prefer popcorn to potato and tortilla chips), it may be more useful to look at the probabilities for various possible values.

If Angus repeated his survey multiple times, each time with another randomly chosen 90-person sample, he could compare results from the surveys, to see how probable it is that 271 is the actual number of students at the school who prefer popcorn, or how close it is to other estimates. The repeated re-sampling method is especially important when trying to draw conclusions about a large target population, such as if Angus wanted to know how many students in all the high schools in the city prefer popcorn.

Here is how you may see proportional reasoning from sample data on the ACT.

Of the 1700 receipts containing large appliance purchases from last month at a certain appliance store, 160 receipts were chosen at random. Each receipt was checked to see which, if any, of the following types of large appliances it included: washer, dryer, refrigerator, or air conditioner. The results were tallied, and the exact percents of receipts that included the large appliances are shown in the diagram below.

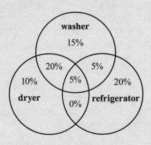

Because this was a random sample, the percents in the sample are estimates for the corresponding percents among all 1700 purchases that included one or more large appliances at the appliance store last month. What estimate does this give for the number of customers who purchased a dryer but none of the other 3 types of large appliances last month?

A. 170
B. 255
C. 340
D. 425
E. 595

Angus recruited some of his friends from other high schools to survey high school students throughout the city as to whether they prefer potato chips, tortilla chips, or popcorn. He instructed each friend to survey multiple randomly-chosen sample groups at his or her high school. Between Angus and his friends, they completed surveys for a total of 100 sample groups of 90 students each, with each sample group randomly selected. The results are shown in the table below, in terms of the frequency (number of sample surveys) with given outcomes (numbers of students who prefer popcorn of the 90 in a given sample survey).

Outcome (Number of Students in Sample Who Prefer Popcorn)	Frequency (Number of Sample Groups)
5	1
19	2
23	3
25	5
27	8
28	9
29	8
30	12
31	11
32	10
33	10
34	8
36	5
38	4
40	3
60	1

According to this table, what is the probability that a survey of a 90-student sample will have an outcome of exactly 33 students who prefer popcorn? What is the probability of a 90-student sample having an outcome of between 28 and 33 students who prefer popcorn? Which survey results might you question the validity of?

Out of the 100 sample surveys, 10 resulted in an outcome of 33 students preferring popcorn, so the percentage of surveys with this result is 10/100 = 0.1. In other words, there is only a 10% chance that exactly 33 students out of 90 will choose popcorn.

Out of the 100 sample surveys, 9 had an outcome of 28, 8 had an outcome of 29, 12 had an outcome of 30, 11 had an outcome of 31, 10 had an outcome of 32, and 10 had an outcome of 33.

$$\frac{9 + 8 + 12 + 11 + 10 + 10}{100} = 60/100 = 0.6$$

There is a 60% chance that between 28 and 33 students in a 90-student sample will choose popcorn. So, Angus's result of 33 students in his initial survey is not unusual. It is within a range of results that occur 60% of the time.

The survey result of only 5 people out of 90 choosing popcorn seems strange, compared to the other results in the table. So does the result of 60 popcorn fans. These may be natural rare occurrences, but they may also be invalid results, perhaps from poorly worded questions in a survey or from a non-random sample, such as the group of students buying snacks at a vending machine that dispenses chips but not popcorn.

Notice that 90% of the sample surveys resulted in outcomes of 25 to 38 students choosing popcorn, per sample. This means a 90% probability that the portion of all high school students in the city who prefer popcorn to potato or tortilla chips is between 25/90 and 38/90.

It is quite a lot of work to coordinate and carry out 100 surveys of 90 people each, with each group chosen through random methods. It also may not be possible to repeatedly re-sample, for various reasons (cost, availability, etc.). In such cases, simulations may allow a researcher to determine the significance of statistical results.

This data presentation is a type of Venn diagram. The percent of customers who purchased only a dryer and no other listed large appliance is shown in the portion of the "dryer" circle that does not overlap with any other circle: 10%. Because 10% of the 160 receipts showed just a dryer purchase, it is likely that 10% of the full population of last month's large appliance purchasers at the store purchased a dryer and none of the other large appliances.

10% of 1700 = 0.1 · 1700 = 170

The data provides an estimate of 170 people who purchased a dryer at the appliance store last month, without purchasing an additional washer, refrigerator, or air conditioner. The correct answer is (A).

The student council consists of 3 representatives from each of the 4 grades at the school. To distribute the responsibilities evenly, the principal is supposed to use a random method to choose which student council representative to contact for each question she has. Over the past year, she has contacted individual representatives with questions a total of 320 times. Geoff was contacted 80 times and suspects that the principal is not using random methods, because his number is so high. What are the theoretical probability and the experimental probabilities that the principal will choose Geoff? Create a simulation model of the principal choosing any of the student council representatives with equal probability (the theoretical probability). Use this to simulate 40 outcomes of the principal randomly choosing a representative. What number of outcomes resulted in Geoff being chosen? What would this translate to, proportionally, out of a total of 320 times?

There are 3 student council representatives from each of the 4 grades, so the total number of representatives is 12. If each representative has an equal chance of being chosen, the probability of the principal choosing Geoff is 1/12. This is the theoretical probability.

Over the past year, the principal chose Geoff 80 out of 320 times, so the experimental probability of her choosing Geoff is 80/320, or 1/4. The experimental probability is 3 times the theoretical probability, so the statistic seems meaningful.

One way to simulate the situation is by rolling a fair 12-sided die (a dodecahedron), because the probability of the die landing with any given numbered face up is 1/12, the theoretical probability of the principal choosing any given representative. Let's assign the number 12 to represent Geoff. Each time the die lands on 12 represents the outcome of the principal contacting Geoff, randomly.

We could choose any number between 1 and 12 to represent Geoff. We just need to choose the number in advance of the simulation, so that we are not somehow influenced by simulation results in our choice of number.

Here are the results of an experiment of rolling a 12-sided die 40 times.

Result	1	2	3	4	5	6	7	8	9	10	11	12
Frequency	4	3	1	0	1	3	8	4	6	2	4	4

In this simulation, the principal chose Geoff 4 out of 40 times, which represents an experimental probability of 1/10, or 10%. Proportionally, this would translate to randomly choosing Geoff $1/10 \cdot 320 = 32$ times.

Notice, however, that this simulation also shows 8 occurrences of the principal randomly choosing representative number 7, which is 1/5 of the 40 simulated results. And, the principal never chose representative number 4 at all in this simulation. As seen here, there is often a great deal of variability in chance processes. The sample size is also not so large. So, although this simulation still suggests that the outcome for Geoff is unusual, it does not clearly prove the unlikeliness of such an outcome.

Another simulation method involves using a random number generator.

EXAMPLE

Suppose Geoff used a random number generator to generate 320 numbers between 1 and 12, counted the number of times 12 appeared in the set, and repeated this two-step process for a total of 10 data sets (each consisting of 320 numbers). This simulates the principal randomly choosing representatives for 3200 questions (i.e., with each representative equally likely to be chosen). Here is the set of results he got.

$$\{24, 28, 30, 15, 32, 31, 17, 29, 30, 25\}$$

Based on this data, is it statistically meaningful that the principal contacted Geoff with 80 of the questions last year?

None of these results is anywhere near 80, the actual number of times the principal sent Geoff her questions. Even the outcome that is furthest from the expected result is not as different. The expected result is $1/12(320) \approx 27$. The outcome of 15 is furthest from that number in this set, but it is not nearly as far away as 80. Based on this data, the actual outcome of 80 seems statistically significant as a very rare occurrence.

This repeated simulation experiment tells us that a random result of Geoff being chosen 80 times is highly unlikely, so it is probable that the principal was not using truly random sampling methods to choose student council representatives to contact.

Lesson 8.4
Probability Distributions

REVIEW

The **median** of a set of data is the middle element, or the mean of the middle two elements for an even number of elements, when all elements in the set are arranged from least to greatest.

A **skewed distribution** is a data distribution that is asymmetrical, with more data toward one end than toward the other. As a distribution curve, it has a longer "tail" on one end than the other, in contrast with a normal distribution, where the curve is symmetrical.

A **uniform distribution** is a data distribution that is completely uniform, where all data values are equal. As a distribution curve, it is a straight horizontal line.

The area of a square of side length s is given by the formula $A = s^2$.

The area of a triangle of base b and height h is given by the formula $A = 1/2\ bh$.

The area of a rectangle of length l and width w is given by the formula $A = lw$.

A **probability distribution** graph shows the probabilities of obtaining various values (outcomes) within that set.

DISCRETE PROBABILITY DISTRIBUTIONS

When the sample space (the set of possible outcomes) is made up of individual, countable outcomes, the probability distribution is a set of points: a **discrete probability distribution**. In the case of a coin toss, where there are only two possible outcomes, the probability distribution is two points: the probability of an outcome of heads and the probability of an outcome of tails.

We may want to assign values (numbers) to outcomes rather than giving them descriptive labels (Heads and Tails, for example), for a probability distribution graph. For the act of tossing a coin once, we can define the number 0 to represent an outcome of tails and the number 1 to represent an outcome of heads. Each outcome has a theoretical probability of 0.5. Here's a graph relating the outcome value to its probability.

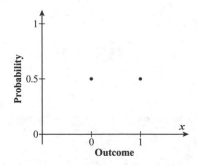

The graph consists of only two points, because there are only two possible outcomes for the coin toss: 0 (tails) and 1 (heads). The two points are at the same level, because they have the same probability: 0.5. This is an example of a **discrete uniform distribution**, a type of probability distribution in which the outcomes are discrete (separate, individual) events and the probability is uniformly distributed (meaning that each outcome has an equal probability).

Another example of a discrete uniform distribution is the probability distribution for a toss of a fair die. The probability is the same (1/6) for each of the 6 possible outcomes (rolling 1, 2, 3, 4, 5, or 6).

The discrete uniform distribution for the coin toss is a representation of theoretical probabilities, not experimental probabilities, of the two possible outcomes. Uniform distributions typically reflect theoretical probabilities, because experimental probabilities do not usually produce exact uniform probability distribution.

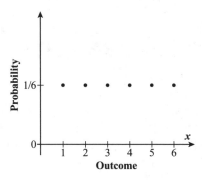

The horizontal axis, labeled here as Outcome, or *x*, is the random variable axis. A random variable represents any of a set of possible values for outcomes of an experiment, or results of a survey or observational study.

If an experiment has exactly *n* different, equally likely individual results, then what will its probability distribution look like?

The outcomes are distinct individual results that are equally likely, so the distribution is a discrete uniform distribution. Because there are exactly *n* possible outcomes, each equally likely, the probability of any one of those outcomes is $1/n$. The discrete uniform distribution for this experiment would be a series of *n* points for random variable integer values 1 through *n*, each at a height of $1/n$ on the probability axis.

Random number generators also use uniform distribution. Typically, a random number generator generates numbers on the (0, 1) interval, each with an equal probability. So, for example, it might break the 1-unit interval into thousandths, assigning each value (0.001, 0.002, and so on) an equal probability, to generate from a group of 1000 distinct numbers.

CONTINUOUS PROBABILITY DISTRIBUTIONS

While a discrete probability sample space contains a countable number of values, a continuous probability sample space contains a range of values, which means it includes the infinite number of points that are within that range.

For an infinite number of possible outcomes, there is an infinite number of probability values associated with them. While the probability distribution graph for a discrete probability distribution is a set of points, the probability distribution graph for a **continuous probability distribution** is a continuous curve.

In a **continuous uniform distribution**, there is equal probability over a given range of values. The sample space range may represent, for example, a range of lengths or areas, each of which would include an infinite number of points.

For example, consider the grape weight data at the beginning of Lesson 8.3. The full set of possible individual grape weights is a continuous distribution of values, with an infinite number of possible values within that range. The range of grape weights includes an infinite number of possibilities because the grape weights can be, theoretically, measured to infinity decimal places.

What is the probability that a randomly chosen point within square *QRST* lies within the shaded triangle *QVT*, as shown below?

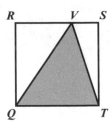

Because all points within *QRST* are equally likely to be chosen (randomly), the probability that the chosen point is in triangle *QVT* is the same for all of them. This means that the probability distribution of being in *QVT*, for the set of randomly chosen points from the square, is a continuous uniform distribution.

There are an infinite number of points in the square and an infinite number of points in the triangle. We must therefore use the ratio of the areas of these two shapes to determine the probability.

Let *s* = the side length of square *QRST*. The area of *QRST* is equal to s^2. Triangle *QVT* has a base that is the same length as one side of the square, *s*, and a height (perpendicular to that base) of the same height as the square, *s*. The area of triangle *QVT* is $1/2 \cdot s \cdot s$, or $1/2\, s^2$.

The ratio of the area of shaded triangle *QVT* to the area of the entire square *QRST* is $\dfrac{\frac{1}{2}s^2}{s^2}$, which simplifies to 1/2. The probability that a randomly chosen point within square *QRST* will lie within triangle *QVT* is 1/2, or 0.5. If the infinite number

of points within *QRST* were assigned individual values and one of those values were

chosen at random, its probability of corresponding to a point within the triangle

would be 0.5.

A range of times also constitutes a continuous sample space with an infinite number of values. If a woman arrives at a certain location at any time (randomly chosen) within a 14-minute range, the probability of her arriving at a specific given time depends on the units in which we break up the time range. For example, there are 14 minutes in the set of possible outcomes, so the probability that she will arrive at any given minute (such as at minute 1, meaning within just the first minute) is 1/14. However, because there are $14 \cdot 60 = 840$ seconds in 14 minutes, the probability that she will arrive at any given second within the 14-minute range is 1/840.

These two probability distributions for the woman's arrival time are both discrete uniform probability distributions. A graph would be a series of 14 points at a probability level of 1/14 for the first approach and a series of 840 points at a probability level of 1/840 for the second approach. But, the actual range of possible arrival times is infinite, matching a continuous uniform probability distribution. We could draw a horizontal line segment to represent the uniform probability for all points in the continuous sample space, but at what level would we draw it?

Instead of using the above approach, with probability on the vertical axis, statisticians instead use the area under a curve to represent the cumulative probabilities for a continuous probability distribution. The total area for the entire sample space must be 1, representing 100%, because all possible outcomes must fall within the sample space range. We were given the fact that the woman arrives sometime within the 14-minute range, so there is a 100% probability that she will arrive somewhere between $x = 0$ and $x = 14$ on the random variable axis. There is 0% probability that she will arrive before 0 or after 14 minutes ($x < 0$ or $x > 14$). The curve (which in the case of a uniform distribution is a straight line segment) is defined by a **probability density function**.

In mathematics, we use the term "curve" to mean any sort of curved or straight line—"line" in the everyday sense of the word, that is. We use the term "line" only to represent a straight line.

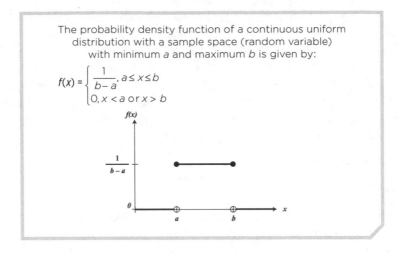

The probability density function of a continuous uniform distribution with a sample space (random variable) with minimum a and maximum b is given by:

$$f(x) = \begin{cases} \dfrac{1}{b-a}, & a \le x \le b \\ 0, & x < a \text{ or } x > b \end{cases}$$

The continuous uniform distribution, shown in the graph above, is also called the rectangular distribution, because the probability is the area under the curve and above 0: a rectangle when the probability density function is horizontal. Probabilities are given for ranges of x-values rather than for individual x-values. However, the results are the same as if we used discrete probability graphs for calculations, with the benefit of only having to use one probability distribution graph for any range of x-values chosen.

For example, let's look again at the probability distribution for a woman arriving any time within a 14-minute range. Our random variable parameters are $a = 0$ and $b = 14$, so the probability function is $f(x) = \dfrac{1}{14 - 0}$, or $f(x) = 1/14$, for $0 \le x \le 14$, and $f(x) = 0$ for $x < 0$ or $x > 14$. Probability distributions often use decimal values for the vertical axis, so let's convert $1/14$ to a decimal: 0.0714. Here is the probability density function for this situation.

Remember, the function curve is just the boundary line for the area underneath, which is what represents the probability. The function value along the curve does **not** give the individual probabilities for values in the sample space.

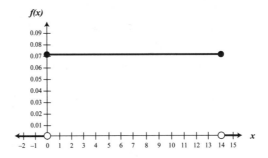

The probability is the area under the curve for the given random variable range. For an arrival time between $x = 0$ and $x = 1$ (or any other one-minute time interval), the probability is the area under $f(x) = 0.0714$ between those two points. This is a rectangle with length 0.0714 and width 1, so the area is 0.0714. This reflects what we found earlier: the probability of the woman arriving within any one-minute period is 1/14, or 0.0714.

The probability that the woman arrives within any given one-second interval is equal to the area under the curve for a one-second width. The x-values on the graph are in terms of minutes, so we must convert one second to minutes. One second is 1/60 minute, or $0.1\overline{6}$ minute. The area under $f(x) = 0.0714$ for a width of $0.1\overline{6}$ is equal to $0.0714 \cdot 0.1\overline{6} = 0.0119$. This is equivalent to 1/840, the probability we found earlier.

EXAMPLE 🔒 14

A 14-minute fountain show continuously repeats, cycling through a 4-minute pop song, a 3-minute excerpt from an opera, a 3-minute Broadway show tune, and a 4-minute classic rock song. If Hailey arrives at some random point in the cycle and watches for 5 minutes, what is the probability that she will *not* see the opera excerpt?

First, let's determine the sample space. In this situation, the sample space is the 14-minute period of time that encompasses one full cycle of the fountain show, because Hailey could arrive at any point in that 14-minute interval. This is a continuous random variable interval.

The "desired" outcome for the probability is Hailey staying for 5 minutes and not seeing the opera excerpt. We must find what range of times, out of the 14-minute interval, fits this description. If Hailey arrives at any point during the 3-minute opera piece, she sees at least some of it. Also, if she arrives less than 5 minutes before the opera piece starts, she sees some portion of it, because she is staying for 5 minutes. So, in total, there are 3 + 5 = 8 minutes out of the cycle with arrival times resulting in Hailey seeing the opera piece.

The arrival points that result in Hailey not seeing the opera piece are along the interval of the remaining 14 – 8 = 6 minutes. So, the probability that Hailey will not see the opera excerpt is 6/14, simplified as 3/7, or about 0.43.

If we are being precise, the number of minutes is less than 8, but for the purposes of this situation, rounding off is fine, especially because the "arrival" time will not be precisely measured to a second. Hailey might also see a few seconds of the opera piece as she's walking up to the fountain, before "arriving."

Now, let's perform the calculation using the probability distribution graph (and the probability distribution function). The area under the function $f(x) = 0.0714$ for a time range of 6 minutes is $0.0714 \cdot 6 \approx 0.43$.

See the histogram of grape weights at the beginning of Lesson 8.3.

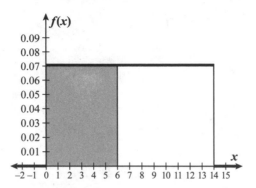

Not all data frequency distributions are uniform, as we saw in the case of grape weights, so not all continuous probability distributions are uniform. If the researcher recorded the individual grape weights instead of categorizing them by intervals, then she would have a set of 10,000 data points. With a very large number of data points, a graph of frequencies of weights would start to resemble a bell-shaped curve. The probability distribution for the outcomes (grape weights) would also resemble a curve, of the same basic shape. A probability density curve for all grape weights (of this variety of grape) may look like the one below.

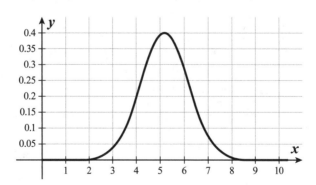

Remember, you cannot read the probability off the y-axis. The probability of a range of x-values is the area under the curve. This curve is symmetrical, so the area under the left half of the curve is equal to the area under the right half of the curve. The midpoint of the curve occurs at x = 5.2. The total area under the curve is 1, so the area under the curve in either half is 0.5. This means there is a 50% chance that a grape of this variety weighs less than 5.2 grams and a 50% chance that a grape of this variety weighs more than 5.2 grams. For a sample data set matching this curve, that would mean about half the grapes weigh less than 5.2 grams and about half the grapes weigh more.

Normal Distribution

Probability distribution curves can take many different shapes, but one of the most common ones is the **normal distribution**. Many biological measurements, such as grape weights or human heights, approximate a normal distribution curve. A normal distribution curve is a type of continuous probability distribution, and its probability density function graph is a symmetrical, bell-shaped curve. Because of its symmetry, the mean and median of the values lie at the same point, aligned with the center of the curve. Half of all values are less than the mean, and half of all values are greater than the mean.

The points of inflection of the curve each lie one standard deviation (σ) from the mean (μ). A data set that is normally distributed but with greater variability is a flatter curve, with a larger σ. A data set that is normally distributed but with less variability is a taller curve that falls off more quickly on either side, with a smaller σ.

The area under each curve is equal to 1. So, a curve with a larger σ will be lower, while a curve with a smaller σ will be higher, to maintain that area.

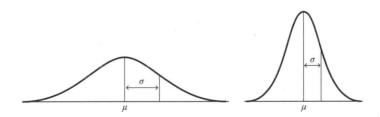

We have seen points of inflection in tangent functions in Lesson 4.4 and in cube root functions in Lesson 6.1. Here, the points of inflection of the bell curve are meaningful as locations one standard deviation away from the mean.

The great thing about normal distributions is that the areas of sections under their curves are known for any given distance from the mean, in terms of standard deviations. If you know the mean and a standard deviation for a normal distribution, you can graph the distribution curve, and you can also calculate the probability for any range of random variable values.

Here are some highlights of the normal distribution relationship:

- About 68% of all observed values are within one standard deviation (1σ) of the mean.

- About 95% of all observed values are within two standard deviations (2σ) of the mean.

- About 99.7% of all observed values are within three standard deviations (3σ) of the mean.

Because the normal distribution curve is symmetric with respect to the mean, these percentages can be divided in half when measuring in one direction from the mean: 34% of all observed values are within σ to the right of the mean, 47.5% are within 2σ to the right of the mean, and 49.85% are within 3σ to the right of the mean. And, it is the same for percentages within 1, 2, and 3 standard deviations to the left of the mean.

This graph indicates the number of standard deviations to the left and right of the mean. The x-value of the point that is labeled as −σ here would actually be μ − σ. The point that is +2σ to the right of the mean would have an x-value of μ + 2σ.

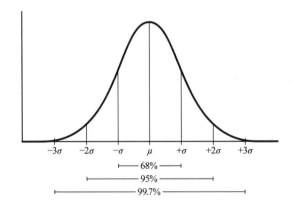

Alexei received a score of 79 on his biology test. He finds out that, for all the students who took that particular biology test, the scores are normally distributed, with a mean score of 70 and a standard deviation of 4. What does that tell him about his test performance in comparison to other students who took the test?

The mean test score (μ) is 70, and the standard deviation (σ) is 4. About 68% of test scores are within one σ, or 4 points, of the mean, 70. So, 68% of students scored between 70 − 4 = 66 and 70 + 4 = 74 on the test.

About 95% of test scores are within two standard deviations, or 8 points, of the mean, so 95% of students scored between 62 and 78 on the test. Alexei's score of 79 is better than the scores of those 95%. Also, because the normal distribution curve is symmetrical, the number of students who scored higher than 78 is the same as the number of students who scored lower than 62. Together, they account for 5% of students who took this test (100% − 95%), so the percentage of students who scored higher than 78 is 1/2 (5%), or 2.5%. This means that Alexei's score is in the top 2.5% of test scores. In other words, he scored higher than 97.5% of students who took that biology test. (He is in the 97.5 percentile.)

This sort of conclusion is only valid if there is a sufficiently large population. If the mean and standard deviation are for a class of 25 students, the data may not match up very well with the normal distribution curve. However, if this particular biology test has been taken by thousands of students, our conclusion is relatively accurate.

EXAMPLE 16

According to historical data, the mean of high temperatures for June 18 in a certain town is 94.5° F, with a standard deviation of 4.5° F. Assuming the data is normally distributed, what is the probability that the temperature in this town will reach a high of at least 90° F next June 18?

The temperature 90° F is 4.5° F less than the mean, 94.5° F, so it is one standard deviation below the mean. The probability of a temperature 90° F or higher is the area under the normal distribution curve from 90° F to the right.

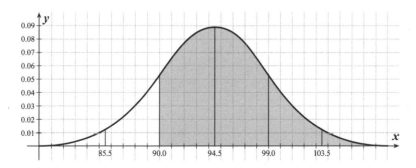

We know that the area to the right of 94.5 contains 50% of the data. The area within one σ on either side of center contains 68% of the data, so the area between 90.0

and 94.5 contains 1/2 (68%) = 34% of the data. The area to the right of 90.0 contains 34% + 50% = 84%, or 0.84. There is an 84% chance that it will reach a high of at least 90° F in this town next June 18.

EXAMPLE 17

Use the term "percentage points" when you are talking about a difference between percentages. Otherwise, saying something like, "a mean of 21.4% and a standard deviation of 2.8%" might be interpreted as meaning that the standard deviation is 2.8% of 21.4%, which is not the case here.

Chris's body fat percentage is 17.0%. Suppose that body fat percentages for men of Chris's age are normally distributed with a mean of 21.4% and a standard deviation of 2.8 percentage points. What does this tell you about where Chris ranks in body fat percentage among men his age?

Chris's body fat percentage is below the mean. We must determine how many standard deviations it is from the mean. In this case, it is not a whole number of standard deviations. First, find the absolute difference in body fat percentage between the mean and Chris's.

21.4 – 17.0 = 4.4

Chris's body fat percentage is 4.4 percentage points less than the mean. Divide this difference by the standard deviation.

4.4/2.8 = 1.57

Chris's body fat percentage is 1.57 standard deviations below the mean. We know that the area to the left of the mean contains 50% of the population and the area between the mean and one standard deviation away contains 34% of the population, so the area to the left of the point ($\mu - \sigma$) contains 0.5 – 0.34 = 0.16.

We could also calculate this area as half (because of symmetry) of the area not included in the 68% within 1σ of the mean: $\frac{100 - 68}{2} = 16$.

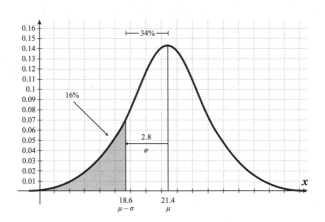

We also know that the area to the left of two standard deviations from the mean contains half of (100 − 95)% of the data (with the other half of that in the right-hand tail).

$$\frac{100 - 95}{2} = 2.5$$

Chris's body fat percentage is between these two, so he is in the bottom 16% of all men his age, but not in the bottom 2.5%, by body fat percentage.

It is possible to calculate the percentile more precisely, using the **z-score**, which is the number of standard deviations an element is from the mean. Here, Chris's z-score is −1.57. Below is an excerpt from a standard normal table, which gives area under the curve to the left of the given z-score.

z	0.09	0.08	0.07	0.06	0.05	0.04	0.03	0.02	0.01	0.00
−1.6	0.0455	0.0465	0.0475	0.0485	0.0495	0.0505	0.0516	0.0526	0.0537	0.0548
−1.5	0.0559	0.0571	0.0582	0.0594	0.0606	0.0618	0.0630	0.0643	0.0655	0.0668

Find −1.5 in the left column and then 0.07 in the top row, to represent the total z-score of −1.57. The box where that row and column intersect shows the number 0.0582. This represents the area under the normal curve, to the left of a z-score of −1.57. To convert to a percent, multiply by 100. This tells us that Chris's body fat percentage is in the 5.8 percentile for men his age. In other words, 5.8% of men his age have a lower body fat percentage than Chris.

EXAMPLE 18

Kuan-Ni is a piano teacher. She has recorded the number of hours per day she has taught for the past few years. Using statistical analysis on her spreadsheet, she found that, on 95% of her work days, she taught between 2.0 and 8.4 hours per day. What is likely the mean number of hours she taught per day? What is likely the standard deviation for the data set of hours per day? What underlying assumption produces the above answers, and could it be wrong?

For a normal distribution, 95% of all data lie within two standard deviations up and down from the mean. If 2.0 is two σ below the mean and 8.4 is two σ above the mean, then the mean is exactly between 2.0 and 8.4.

$$\mu = \frac{2.0 + 8.4}{2} = 5.2$$

There is a distance of four standard deviations from 2.0 to 8.4. Divide the difference by 4 to solve for σ.

$$\sigma = \frac{8.4 - 2.0}{4} = 1.6$$

So, Kuan-Ni taught an average (mean) of 5.2 hours per day, with a standard deviation of 1.6 hours for her daily teaching hours data.

To produce these answers, we assumed that Kuan-Ni's set of daily teaching hours is normally distributed. This may not be true. For example, even though 95% of her teaching days include 2.0 to 8.4 hours per day, her mean could be 6.3 hours per day, with her median even higher. The distribution could be skewed left, as shown below.

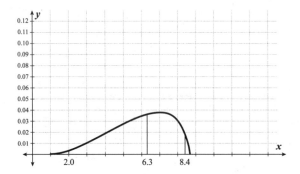

As part of their marketing campaign, a store sent out E-mails containing coupons to all 1800 people on their mailing list. For each group of 100 E-mail addresses, they used a computer program to randomly sort the addresses into two groups of 50 addresses each. To the addresses in Group A, they sent a "buy one, get one 1/2 off" coupon, which gives the customer a 50% discount on one item with the purchase of another item at full price. To the addresses in Group B, they sent a "25% off" coupon, which gives the customer a 25% discount off the total price of up to two items. The company tracked the number of coupons of each kind used by members of each group of 50, as shown below.

"Buy one, get one 1/2 off" coupon:
{ 9, 10, 10, 11, 11, 11, 12, 12, 12, 12, 12, 12, 12, 13, 13, 14, 15, 15}

"25% off" coupon:
{17, 17, 18, 18, 18, 19, 19, 19, 19, 19, 19, 19, 19, 20, 20, 20, 20, 22}

(a) Which of the two coupons seems to result in more customers using the coupon? Use the means and standard deviations of the two sets to explain your answer.

(b) The marketing department combined the results for each group of 100 mailings and randomly divided each into a group of 50, then determined the number of coupons (either kind) used within that group. They calculated the mean number of coupons used per 50 customers receiving *either* of the E-mails, as well as the standard deviation for this combined set of data. They used a computer software program that generates random numbers according to a normal distribution, with the mean number of coupons (either type) used per 50 customers, and the standard deviation for that set. In five simulations, they got the following data sets.

{11, 13, 14, 15, 15, 15, 15, 16, 16, 17, 17, 18, 20, 20, 21, 21, 23, 23}
{5, 5, 11, 12, 13, 13, 14, 14, 15, 16, 16, 16, 16, 17, 19, 19, 24, 24}
{11, 12, 12, 13, 14, 14, 15, 18, 18, 18, 18, 19, 19, 19, 19, 19, 20, 21}
{9, 9, 9, 10, 10, 11, 12, 13, 13, 13, 14, 14, 15, 16, 17, 18, 18, 22}
{7, 8, 9, 10, 14, 14, 14, 14, 14, 15, 15, 15, 16, 16, 16, 17, 20, 22}

According to this data, how likely is it that the "25% off" coupon results are due to randomness in how the treatment sets were assigned? What are possible explanations for the difference in the numbers of coupons used?

It seems that the marketing department wants to determine which wording has a greater effect on customers. Each of these coupons results in a maximum discount of 25% on a purchase of two items.

The "25% off" coupon seems to result in more customers using the coupon. The mean coupons used (per 50-address mailing) is 19 for this set, as compared with a mean of 12 coupons used for the "buy one, get one 1/2 off" set. The variability is pretty low for each set (1.6 for the "buy one, get one 1/2 off" group and 1.2 for the "25% off" group), which means that data is pretty consistent within each set, and there is no overlap between the sets.

None of the simulation sets seems very similar to either of the single-coupon-variety data sets. The third simulation set has somewhat similar numbers, in terms of the median and the upper half of the data, to the "25% off" coupon data set, but the range is much greater, and the standard deviation is much greater (3.2, as compared to 1.2).

It seems very unlikely that the greater use of the "25% off" coupons is due to randomness in how the treatment sets were assigned. One possibility is that "25% off" sounds like a better discount to customers than "buy one, get one 1/2 off." Another possibility is that the difference is due to some customers using the "25% off" coupon to purchase just one item, whereas they would not have used it to purchase two items. This study did not take this factor into account.

Lesson 8.5
Sample Proportions and Sampling Distribution

A **Bernoulli trial** is a random experiment that has exactly two possible outcomes in its sample space (called "successes" and "failures"), with the same probability of success for every time the experiment is repeated. A repeated coin toss is one example of a Bernoulli trial. Repeatedly drawing a marble from a bag of 7 purple and 3 green marbles, with replacement of the drawn marble each time, is also a Bernoulli trial.

If we wanted to determine the percentage of voters who will vote to re-elect the current president, we might poll a sample of the population, because we cannot poll every voter in the country. A sample proportion is the ratio of successes to total number of experiments or observations in the sample. If we polled 200 voters, and 72 of them said they would vote for the incumbent president, then the sample proportion is 72/200, or 0.36.

If you did not put the marble back in the bag after drawing, the total number of marbles in the bag and the number of marbles of one color would both decrease, resulting in a different probability for the next draw.

> **Sample proportion**, $\hat{p}$, is given by $\hat{p} = x/n$, where x = the number of successes and n = the number of Bernoulli trials or number of observations in the sample.

For large values of n (large sample sizes), the sample proportions of Bernoulli trials follow an approximately normal distribution. A good rule of thumb is that the distribution of $\hat{p}$ will be normal if the sample is large enough to include at least 10 successes and at least 10 failures.

Ideally, these numbers of successes and failures are based on the "true" percentages (population proportions) applied to the sample size. However, in the case that the population proportion, p, is unknown, we use the sample proportion, $\hat{p}$, i.e., the actual number of successes and failures in the sample.

> If samples are repeatedly drawn from a population (or a Bernoulli trial is repeatedly performed), with $np \geq 10$ and $n(p - 1) \geq 10$, then the sample proportions are approximately normally distributed.

The idea is that there is some underlying "true" percentage (the percentage of total voters who will vote to re-elect the incumbent president; the 50% probability of a coin landing on heads), and repeated trial/sample percentages will be normally distributed in relation to this percentage. In other words, $p = \mu$ for the normal distribution curve of sample proportions.

> When the population size is at least 20 times as large as the sample size, then the standard deviation of the **sampling distribution** (a normal distribution of sample proportions) is given by the formula $\sigma_{\hat{p}} = \sqrt{\dfrac{p(1 - p)}{n}}$, where p is the population proportion and n is the sample size.

Patra took restaurant rating data from a website that allows users to rate restaurants with a star system, with the lowest rating being one star and the highest rating being five stars. She took randomly chosen samples of 20 review ratings for each of seven restaurants and calculated the mean and standard deviation for each data set, as shown in the table below.

Restaurant	A	B	C	D	E	F	G
Mean Number of Stars Received	2.6	3.3	2.3	3.2	3.1	2.2	3.9
Standard Deviation, in Stars	1.6	1.7	1.6	1.8	1.8	1.4	1.3

She then took random samplings of seven sets of 20 reviews from the full set of 5000 reviews on the website. The means and standard deviations of these sets are shown below.

Sample of 20 Review Ratings	S_1	S_2	S_3	S_4	S_5	S_6	S_7
Mean Number of Stars Received	2.8	3.0	3.2	3.1	3.0	3.5	3.4
Standard Deviation, in Stars	1.4	1.6	1.3	1.5	1.5	1.6	1.5

(a) If a restaurant received the same number of each level of review within a set of 20 reviews, what would be the mean and standard deviation for the number of stars received per review? How is Patra's data different?

(b) Based on a theoretical probability of each star rating being equally likely, what would be the expected number of times a rating of 3 stars would appear in any 20-review sample set? What would be the standard deviation for the sampling distribution of 3-star ratings, assuming normal distribution? Suppose Patra recorded the number of 3-star ratings in each of her 14 data sets, as follows: {0, 0, 0, 1, 1, 1, 1, 1, 1, 2, 2, 2, 3, 3}. What might the probability distribution graph for actual ratings, in stars, look like? What is a possible explanation for this difference in sampling distribution?

(a) If a restaurant received the same number of each level of review within a set of 20 reviews, then it would receive the following numbers of stars: {1, 1, 1, 1, 2, 2, 2, 2, 3, 3, 3, 3, 4, 4, 4, 4, 5, 5, 5, 5}. The mean for this set is 3, and the standard deviation is 1.45. Patra's sets have similar means (the average of all the means is about 3), but higher standard deviations in most cases. This means there is more variability in Patra's data.

(b) If each star rating were equally likely, then a rating of 3 stars would have a probability of 1/5. In a 20-review sample set, there would be about 1/5 (20) = 4 reviews of 3 stars.

1/5 (5000) ≥ 10 and 4/5 (5000) ≥ 10, so we can determine a normal distribution

based on this theoretical probability. The standard deviation of the sampling

distribution (in terms of sample proportions) would be $\sqrt{\dfrac{\frac{1}{5}\left(1-\frac{1}{5}\right)}{5000}} \approx 0.0057$. So,

the probability distribution for a result of 3 stars would have a mean of 1/5 and a

σ of 0.0057. We can convert this to number of 3-star ratings per 20-rating sample

by multiplying by 20. The probability distribution for number of 3-star ratings per

20-rating set would have a mean of 4 reviews and a standard deviation of 0.11

review.

The actual number of 3-star ratings per 20-rating set are much lower. All of them are fewer than 4 reviews. With a σ of 0.11 review, the theoretical probability set for number of 3-star ratings per sample set should have 99.7% of outcomes within 3(0.11) review of the mean. In other words, 99.7% of outcomes should include between 3.67 and 4.33 stars. The actual number of 3-star ratings consistently fall in the other 0.3% probability space. This means that Patra's results are not at all consistent with a normal distribution of ratings.

With very few 3-star ratings but a mean of around 3 stars per review, and a large standard deviation, Patra's data consists of mostly very high ratings and very low ratings, so it has a bimodal probability distribution. The actual probability distribution of individual ratings, in terms of number of stars, is a discrete probability set and might look something like this.

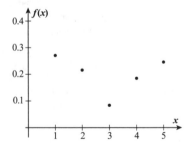

One possible explanation for the star rating data is that there is a voluntary bias. The ratings are given voluntarily, and the people who post those ratings on the website tend to be people who feel very strongly about a restaurant, whether positively or negatively. People who consider the restaurant average (3 stars) may not be as inspired to post a review.

Elton and Robin are playing a game in which they roll two dice and record the sum. Out of 100 sums recorded, there were 20 sums of 10 or greater. Elton thinks this is unusually high. Use the sampling distribution for this series of Bernoulli trials to determine the probability of rolling a sum of 10 or greater 20 out of 100 times.

Because the probabilities of dice follow uniform distributions, we can find the true population proportion: the theoretical probability of rolling a sum of 10 or greater. There are a total of $6 \cdot 6 = 36$ possible outcomes for rolling two dice, and 6 of those outcomes result in a sum of 10 or greater $(4 + 6, 5 + 5, 5 + 6, 6 + 4, 6 + 5,$ and $6 + 6)$, so the theoretical probability of rolling a sum greater than or equal to 10 is 6/36, or 1/6.

In each sum, the first number represents the value on die number 1 and the second number represents the value on die number 2. We must count both $4 + 6$ and $6 + 4$ as distinct ways of rolling a sum of 10.

Let's test if this situation is eligible for fitting to a normal distribution curve.

$np = 100 \cdot 1/6 = 16.\overline{6}$
$n(1 - p) = 100 \cdot 5/6 = 83.\overline{3}$

Both of these projected success and failure numbers are greater than 10, so the distribution of the sample proportions is approximately normal.

Using $p = 1/6$ and $n = 100$, we can calculate the standard deviation for the sampling distribution.

$$\sigma_{\hat{p}} = \sqrt{\frac{\frac{1}{6}\left(1 - \frac{1}{6}\right)}{100}} = \sqrt{\frac{\frac{1}{6} \cdot \frac{5}{6}}{100}} = \frac{\sqrt{5}}{60} \approx 0.037$$

The population proportion, or theoretical probability, of 1/6, or $0.1\overline{6}$, is the mean for the sampling distribution, and the standard deviation is 0.037.

The mean for the sampling distribution is also called the **expected value**. In this trial, you theoretically expect to have a 1/6 success rate.

The sample proportion (the experimental probability) in this case is 20/100, or 1/5, because Elton and Robin rolled a sum of at least 10 in 20 of their 100 tries. We want to compare this $\hat{p}$ of 1/5, or 0.2, to the mean of $0.1\overline{6}$, in terms of a standard deviation of 0.037.

The difference between $\hat{p}$ and p is $0.2 - 0.1\overline{6} = 0.0\overline{3}$. This is an absolute difference. We need to convert it to be in terms of numbers of standard deviations.

$0.0\overline{3}/0.037 \approx 0.9$

The sample proportion, 0.2, is 0.9 of a standard deviation to the right of the mean, $0.1\overline{6}$.

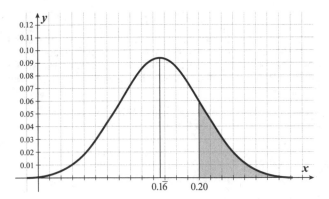

Notice that, because our random variable changed, the standard deviation also changed. For the distribution of trial success numbers, the standard deviation is 3.7.

The value 1σ to the right of the mean is $0.1\overline{6} + 0.037 = 0.204$. The percentage of area under the curve to the right of this value is $1 - 0.84 = 0.16$. This means that in 16% of trials (each trial a set of 100 rolls of the dice), the proportions of successes (getting a sum of 10 or greater) to number of dice throws is 0.204 or more. Elton and Robin's sample proportion, 0.2, is to the left of 0.204, so the above statement definitely also applies to a sample proportion of 0.2. In other words, you can expect a 1/5 success rate in about 16% of 100-toss trials.

It might be easier to interpret a distribution of trial results (number of times a sum of 10 or greater is rolled per trial) instead of a distribution of sample proportions. The number of expected successes, according to the theoretical probability, is np: $100 \cdot 1/6 = 16.\overline{6}$. The actual number of successes for sample proportion $\hat{p}$ is $n\hat{p}$. It is the $n\hat{p}$ that follows a normal distribution with a mean of np in the graph below, showing probability distribution of sample/trial results.

Because the sample size is 100 for each trial, the number of successful trial results looks very similar to the sample proportion (percentage) for successful trial results. Multiplying each proportion value by 100 simply moves the decimal point two places to the right.

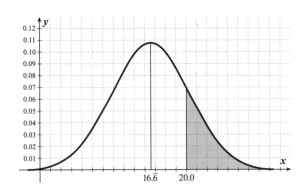

This graph makes it more apparent that a result of 20 or more successes occurs in 16% of trials (each trial being 100 tosses of two dice). This is not such an unusual occurrence.

A typical cutoff for determining a result to be highly unusual is subjective but is often when that result occurs only 5% of the time or less, under a normal distribution.

Lesson 8.6
Confidence Intervals and Margins of Error

In many studies and experiments, the true population proportion, p, and standard deviation, σ, are unknown. In these cases, we use the sample proportion, $\hat{p}$, to approximate p. We also use the same formula we would for calculating standard deviation but substitute $\hat{p}$ for p. This sample standard deviation is called the standard error.

> The **standard error** for a sample proportion distribution is given by the formula SE = $\sqrt{\dfrac{\hat{p}(1 - \hat{p})}{n}}$, where $\hat{p}$ is the sample proportion and n is the sample size.

A standard error functions like a standard deviation in interpretations of normal data distributions. For example, 68% of data lies within one SE of the sample mean, 95% lies within two SE, and 99.7% lies within three SE. However, because we do not know the true population proportion, p, these percentages instead tell us how likely p is to fall within these ranges of our $\hat{p}$ value. In terms of quantities (np or $n\hat{p}$) instead of proportions, we use the standard error to determine whether the true mean of the number of successes is likely to fall within certain ranges of the sample mean, $\mu_{\hat{p}}$, of numbers of successes. These ranges are called confidence intervals.

Compare this to the formula for the standard deviation of the sampling distribution, shown in Lesson 8.5.

> For our purposes, a **confidence interval** is an interval that is likely to contain the population mean, to the specified level of confidence. The most common confidence levels used are 0.90, 0.95, and 0.99.

A confidence interval is an interval that is likely, to the specified confidence level, to contain some population parameter of interest. This could be some other parameter, such as the standard deviation, rather than the mean. Here, however, we will only look at confidence intervals for population means.

The correlation of ±2 standard errors or ±2 standard deviations to 95% is a rounded value. To be more precise, 95% of the population is within 1.96σ or $1.96SE$ of the mean.

> The **margin of error** is the distance from the sample mean that is likely to include the population mean, to the desired confidence level. So, the margin of error is half the width of the confidence interval for a given confidence level.

The margin of error is often expressed using the ± symbol. For example, a report may say, "The liquid's pH is 5.21 ± 0.05," to indicate that the pH measure is very likely in the range between 5.16 and 5.26.

EXAMPLE 🔒22

Dunja wants to estimate the number of people who will vote for the Democratic candidate for mayor in her town's next election. She surveys a random sample of 180 people, 117 of whom said that they would vote for the Democratic candidate. There are typically about 20,000 people who vote in the town's mayoral elections. What is the confidence interval for the number of votes the Democratic candidate will actually receive, at a 68% confidence level? What is the confidence interval for the number of votes the Democratic candidate will actually receive, at a 95% confidence level? Express each of these using the margin of error. If Dunja surveyed an additional 320 people, with the same sample proportion result, how would that affect the margin of error for a 95% confidence level?

The true proportion of people who will vote for the Democratic candidate is unknown, but the sample proportion is 117/180, or 0.65. Clearly, $0.65(20,000) \geq 10$ and $(1 - 0.65)(20,000) \geq 10$, so we can use a normal distribution model.

We are using rounded values, because the SE cannot be calculated with much precision for a small sample size.

$$SE = \sqrt{\frac{\hat{p}(1 - \hat{p})}{n}} = \sqrt{\frac{0.65(1 - 0.65)}{500}} \approx 0.036$$

The sample mean, $\mu_{\hat{p}}$, is 0.65, and the standard error, SE, is 0.036. A total of 68% of possible success ratios (percents of people voting for the Democratic candidate) should be within one SE of the sample mean.

0.65 − 0.036 = 0.614
0.65 + 0.036 = 0.686

For a 68% confidence level, the confidence interval is from 0.614 to 0.686. In other words, there is a 68% probability that the Democratic candidate will receive between 61.4 and 68.6 percent of the votes. The margin of error for the proportion is one SE, or 0.036. So, for a 68% confidence level, the Democratic candidate will receive 65% of the votes plus or minus 3.6 percentage points.

A total of 95% of possible voting ratios should be within 1.96SE of the mean.

$$0.65 - 1.96(0.036) \approx 0.579$$
$$0.65 + 1.96(0.036) \approx 0.721$$

For a 95% confidence level, the confidence interval is from 0.579 to 0.721. In other words, there is a 95% probability that the Democratic candidate will receive between 57.9 and 72.1 percent of the votes. The margin of error, in this case, is $1.96(0.36) \approx 0.071$. So, for a 95% confidence level, the Democratic candidate will receive 65% of the votes plus or minus 7.1 percentage points.

If Dunja surveyed an additional 320 people, with the same proportion result of 0.65, then she will have surveyed a total of 180 + 320 = 500 people, 65% of whom said that they would vote for the Democratic candidate. We must recalculate the SE for this adjusted sample size.

$$SE = \sqrt{\frac{0.65(1-0.65)}{500}} \approx 0.021$$

For a 95% confidence level, the margin of error is now $1.96(0.021) \approx 0.041$. The margin of error has decreased, from about 0.07 for the 180-person sample, to about 0.04 for the 500-person sample. For this 500-person sample, there is a 95% chance that the Democratic candidate will receive 65% of the votes plus or minus 4.1 percentage points.

The margin of error is expressed in the units of the standard variable. Because we define the sample mean as 65 percent, the standard error is in percentage points. Just as we multiplied 0.65 by 100 to convert it to a percent, we must also multiply 0.036 by 100.

Here is how you may see margins of error on the SAT.

Bit Bikes Inc. conducted a random survey of 3,500 bike owners from around the United States. 420 of the respondents stated that they own youth bikes, while the remainder stated that they own adult bikes. After analyzing the results, Bit Bikes Inc. determines that the survey results have a margin of error of 2%. Which of the following best represents the range for the percentage of the nation's bike owners who own adult bikes?

A) 10–14%

B) 12–16%

C) 81–85%

D) 86–90%

If Dunja wanted to estimate the actual number of people who will vote for the Democratic candidate, rather than what percent will vote for him, she would apply all percentages to the actual population of 20,000 voters.

According to her 180-person sample, there is a 68% probability that the Democratic candidate will receive 13,000 votes ± 720 votes. According to the same sample, there is a 95% probability that the Democratic candidate will receive 13,000 votes ± 1420 votes. According to the complete 500-person survey, there is a 95% probability that the Democratic candidate will receive 13,000 votes ± 820 votes.

> For a given data set, as the confidence level increases, the corresponding confidence interval also must increase, which means that the margin of error increases. Viewed in the other direction, a decrease in the margin of error requires a decrease in the confidence level for the data set.
>
> However, an increase in sample size will decrease the margin of error without reducing the confidence level, provided that outcomes for the larger sample remain relatively consistent with the original sample.

EXAMPLE 23

Dave is concerned that he may have a certain medical condition, which exists in 3% of the population. He takes a diagnostics test that indicates that he does have the condition. For people who truly have the condition, this diagnostics test correctly diagnoses them with an 82% confidence level, meaning that it produces false negative results 18% of the time. For people who do not actually have the condition, this diagnostics test correctly diagnoses them with a 90% confidence level, meaning that it produces false positive results 10% of the time.

The probability, based on his test results, that Dave truly does have the condition is the ratio of true positive results to total positive results (both true and false) that the diagnostics test would produce for the population.

Treatment for this condition is expensive and involves potentially serious side effects, so Dave only wants to get treatment if his probability of actually having the condition is greater than 25%. Should Dave seek treatment?

If we define p to be the total number of people in the population, then the number of people who actually have the medical condition is $0.03p$ (3% of p). Of these $0.03p$ people, 82% would be correctly diagnosed using this test, for a total of $0.82(0.03p) = 0.0246p$ true positive results.

Because 3% of the population has the condition, the other 97% does not. The number of people who do not have the condition is $0.97p$. Of these people, 10% will get incorrectly diagnosed as having the condition. This means a total of $0.1(0.97p) = 0.097p$ false positive results.

$$\text{Probability Dave has the condition} = \frac{\text{number of true positive results}}{\text{number of true positive results} + \text{number of false positive results}}$$

$$= \frac{0.0246p}{0.0246p + 0.097p}$$

$$= \frac{0.0246p}{0.1216p}$$

$$= 0.0246/0.1216$$

$$\approx 0.20$$

The probability that Dave actually has this medical condition is about 20%, which is less than 25%. According to his conditional statement, he should not seek treatment.

According to survey results, 420 of 3500 bike owners own youth bikes, so $3500 - 420 = 3080$ own adult bikes. As a percentage of bike owners surveyed, this is $3080/3500 = 0.88$, or 88%. Based on these survey results, about 88% of bike owners in the United States own adult bikes.

The margin of error is 2 percentage points, so the actual percentage of bike owners in the United States who own adult bikes is highly likely be between $88 - 2$ and $88 + 2$ percent, or between 86% and 90%. The correct answer is (D).

CHAPTER 8 PRACTICE QUESTIONS

Directions: Complete the following open-ended problems as specified by each question stem. For extra practice after answering each question, try using an alternative method to solve the problem or check your work.

1. A school has to eliminate some of its extracurricular activities. It surveys a randomly selected group of 100 of the 2500 students in the school, providing options of four extracurricular activities and allowing each student to choose just one to nominate for elimination. The survey finds that of the 100 students surveyed, 42 think that band should be eliminated, 24 think that chorus should be eliminated, 18 think that the golf team should be eliminated, and 16 think that the cooking club should be eliminated.

   (a) Given this survey, how many students in the school likely believe that the school should eliminate each of the four activities?

   (b) The principal decides to eliminate three out of the four extracurricular activities mentioned on the survey. Based on the survey results, she concludes that most students in the school want to keep the cooking club. What is the flaw in her reasoning? How should she redesign the survey if she wants to choose three activities to eliminate based on the one activity the most students want to keep?

2. A bowler in a tournament bowls 8 games and gets scores of 190, 255, 210, 160, 173, 188, 206, and 224.

   (a) Determine the mean, range, and interquartile range of her scores. Also calculate the variance and standard deviation for the set.

   (b) Suppose that this is a representative sample of the 250 games this bowler has played this year. What is the estimated standard deviation of her scores for the year? A total of 68% of her scores are likely to fall within what range of scores? Express this as a number of games.

3. A student is conducting a survey for her Psychology class. She has heard that there is a correlation between the color of a jersey a team wears and the perceived aggressiveness of the team; for example, a team that wears black jerseys is perceived as more aggressive than one wearing white jerseys.

   (a) She wants to conduct a survey in her school to see who agrees with this notion. Her first thought is to just ask everyone in all of her classes, giving a survey to each person that shares a class with her. Her second idea is to ask her teachers to give them out to all students in their classes each day. What is wrong with each of these methods, and what method could she use instead to ensure she gets survey responses from a representative sample of schoolmates?

   (b) What are the problems with using a survey to study this correlation? How might a scientist design an experiment to study the correlation with less bias?

4. Describe the type of probability distribution for each of the following:

   (a) the theoretical probabilities of Lillian choosing each of her seven scarves if she randomly chooses one to wear

   (b) the set of possible experimental probabilities of Lillian choosing her blue scarf if randomly choosing one to wear each day

   (c) the probabilities that an adult alligator chosen at random will measure any given length

   (d) the probabilities that an alligator chosen at random from a group of 50 adults and 50 one-week-olds (not previously measured) will measure any given length

5. A movie theater surveys its patrons regarding its pricing for food and beverage. Of the 1200 people who visited the theater on a given day, 1056 of them said they would not pay more than $15.00 for a combo that included a large drink and a large popcorn. The theater determines that there is a margin of error of 18 patrons for a 68% confidence level for an extrapolation of these survey results to the full population of their patrons. For a 95% confidence level, what is the likely percentage range for the moviegoers who would not purchase the combo at a price more than $15.00?

6. A teacher sets her scoring curve along a normal distribution curve based on the results of a given test. On this particular test, 68% of the students scored between a 56 and an 82. In order to pass the test, the teacher requires a student to have achieved more than the mean score. In order to get an A, the student must be in the 95th percentile, and to get an A+ on the test, a student must achieve a score greater than three standard deviations above the mean. What score would be needed to pass? To get an A? To get an A+?

7. A popular television show ran for 10 seasons, with a total of 236 episodes. Now, a certain channel shows seemingly randomly chosen episodes of this show as reruns, with 2 episodes per night, 5 nights per week, for a total of 10 episodes per week. In the past 20 weeks, there has never been a repeat of the same episode within a given week. (For our purposes, define a given week as Sunday through Saturday.) Based on this information, is the selection of episodes likely to be truly random? Use statistics to support your conclusion.

8. A factory uses quality control tests for flashlights they produce. The tests correctly identify defective flashlights 80% of the time and correctly identify flashlights in perfect working condition 90% of the time. Long-term data suggests that 5% of all flashlights this factory produces are actually defective. The factory discards all flashlights labeled as defective by their quality control tests and sells the rest.

(a) Out of a batch of 10,000 flashlights, how many will the factory discard? What is the probability that a consumer purchasing a flashlight made in this factory gets a defective one?

(b) The factory makes a profit of $12 per flashlight, so they lose this amount of potential profit per properly working flashlight that they discard. The CEO is considering improving the quality control tests. For a cost of $8,000 per batch of 10,000 flashlights, the improved tests would cut the number of false positive and false negative results each in half. Is it worth it to pay for the improved quality control tests, from a cost perspective?

SOLUTIONS TO CHAPTER 8 PRACTICE QUESTIONS

1. **(a) Band: 1050, Chorus: 600, Golf: 450, Cooking Club: 400; (b) See reasoning below.**
    (a) Set up a proportion using the sample data. Of the students surveyed, 42/100 believe band should be eliminated, 24/100 believe chorus should be eliminated, 18/100 think the golf team should be eliminated, and 16/100 say the cooking club should be eliminated. Set each of these fractions equal to an unknown number x over 2500, the actual number of students in the school. Cross-multiply and solve for x in each of the resulting proportions.

    Band: $42/100 = x/2500$, so $x = 1050$ people
    Chorus: $24/100 = x/2500$, so $x = 600$ people.
    Golf: $18/100 = x/2500$, so $x = 450$ people.
    Cooking Club: $16/100 = x/2500$, so $x = 400$ people.

    Therefore, based on this sample, 1050 people would believe band should be eliminated, 600 believe chorus should be eliminated, 450 believe the golf team should be eliminated, and 400 believe the cooking club should be eliminated.

    (b) The principal assumes that the extracurricular activity that the fewest students vote to eliminate is the one that the most students want to preserve, which is not necessarily the case. There are four different activities to choose from, and some may be more polarizing than others in terms of how students feel about them. It's possible that the majority of the 84 students who voted to eliminate one of the other three activities besides the cooking club would also prefer to preserve a different one of those three activities, over the cooking club. The principal should redesign the survey to ask which extracurricular activity each student would most like to preserve and, based on the results, keep the most popular and eliminate the other three activities.

2. **(a) Mean: 200.75, Range: 95, Interquartile Range: 36.5, Variance: 788.1875, Standard Deviation: 28.07; (b) Estimated Standard Deviation: 30.01, 68% of Scores Within 171 and 231, for 170 games.**
    (a) First, calculate the mean by adding all the scores together and dividing by the 8 games bowled. The total would be $190 + 255 + 210 + 160 + 173 + 188 + 206 + 224 = 1606$ total pins, so the mean would be $1606/800 = 200.75$. The range is the difference between the highest and lowest scores, so that would be $255 - 160 = 95$. For the interquartile range, first reorder the numbers; the new order, from least to greatest, would be 160, 173, 188, 190, 206, 210, 224, and 255. The interquartile range is the difference between the upper quartile and lower quartile of the data set. The upper quartile is the median of the upper half of the set, {206, 210, 224, 255}, or $\dfrac{210 + 224}{2} = 217$. The lower quartile is the median of the lower half of the set, {160, 173, 188, 190}, or $\dfrac{173 + 188}{2} = 180.5$. The interquartile range is $217 - 180.5 = 36.5$.

To calculate the variance, determine how much each of the given scores deviates from the mean using the variance equation. The result would be the following:

$$\frac{(160 - 200.75)^2 + (173 - 200.75)^2 + (188 - 200.75)^2 + (190 - 200.75)^2 + (206 - 200.75)^2 + (210 - 200.75)^2 + (224 - 200.75)^2 + (255 - 200.75)^2}{8}$$

This simplifies as 6305.5/8, or 788.1875, so that is the variance. The standard deviation would be the square root of this number, which is 28.07.

(b) If these 8 scores are a representative sample of a total of 250 scores, then we must calculate the estimated standard deviation for the 250 scores, based on this sample. The formula for estimated standard deviation is the same as for standard deviation, except that the sum of squared differences is divided by $(n - 1)$ instead of n.

$$\sqrt{\frac{6305.5}{8 - 1}} \approx 30.01$$

Assuming the bowler's scores follow a normal distribution, 68% of these scores will be within one standard deviation of her mean score. We must use the estimated standard deviation, rounded off to 30, and the mean, rounded off to 201. (Bowling scores are whole numbers.)

201 – 30 = 171
201 + 30 = 231

So, 68% of her scores are likely to be between 171 and 231. Out of 250 total games, 68% is 0.68(250) = 170. Out of the bowler's 250 games played this year, about 170 games should have scores between 171 and 231.

3. **See explanations below.**
   (a) If she only surveys those in her class, the results may be skewed based on shared characteristics among those in her own grade level, with her own academic interests, and/or at the same academic level (such as in honors classes). If she asks only her own teachers to give the surveys out to all students in their classes, likewise, the results will be skewed. If she is trying to conduct a survey in the school to see who agrees with the notion, she needs to ensure she surveys a representative and random population of students in the school. Therefore, she could, for example, give the survey out at lunch to random people in the cafeteria. If she wants the teachers to help, she could also pick a subject that all students must take and give a stack of the surveys to each teacher for that subject —for example, give every English teacher 10 surveys to have students fill out. Every teacher could distribute the surveys randomly to students in one of their classes, ensuring that a solid random cross-section of the school as a whole is surveyed.

   (b) A survey asks people to recognize and represent their own perceptions, but they may not be consciously aware of how color affects their perceptions of aggressiveness. This introduces a bias. To study the correlation with less bias, a scientist could edit a video of two teams playing a game, so that the jerseys for one team appear as all black, all white, or all another color in each of the edited versions of the video. These versions of the same video clip, with different color jerseys, could be shown to a large sample population, with each subject asked to rate the aggressiveness of the given team on a scale of 1 to 10. The results will show whether the same team, behaving the same way, is perceived as more or less aggressive based only on the color of their jerseys, the isolated factor in the experiment.

4. **(a) 1/7, (b) 0 to 1, (c) a continuous normal distribution, (d) bimodal probability distribution**

   (a)   The theoretical probability of Lillian choosing any one of her seven scarves is 1/7, because they each have an equal chance of being randomly chosen. Each probability is the same, and there are exactly 7 distinct options, so this is an example of a discrete uniform probability distribution.

   (b)   Possible experimental probabilities of Lillian randomly choosing her blue scarf range from 0 to 1, but these experimental probabilities are not equally likely. Experimental probabilities close to 1/7 are likely, while experimental probabilities close to 0 or 1 are less likely. The experimental probabilities should follow a continuous normal distribution curve with a mean of 1/7, especially for a large sample space (many samples of sets of days of choosing one scarf each day). For a smaller sample space, there will be a greater standard deviation, which may result in a skewed distribution, because 1/7 is much closer to 0 than to 1.

   (c)   Lengths of adult alligators are probably normally distributed, because many biological measurements are. So, the probability distribution curve for weights of an adult alligator chosen at random should be a continuous normal distribution.

   (d)   In a group of 50 adult alligators and 50 one-week-old alligators, the frequencies of lengths should be highest around the average length of an adult alligator and the average length of a one-week-old alligator. These lengths are very different. So, the probabilities of lengths for an alligator chosen at random from such a group of 100 alligators follow a bimodal probability distribution.

5. **85 to 91%**

   The margin of error is 18 for a 68% confidence level, so there is a 68% probability that between 1056 − 18 and 1056 + 18 out of 1200 moviegoers are unwilling to pay more than $15.00 for the combo. The margin of error of 18 is like a standard error, or standard deviation, for a normal distribution. To find the margin of error for a 95% confidence level, we must find 1.96 times 18.

   $1.96 \cdot 18 = 35.28$

   Out of 1200 moviegoers, there is a 95% chance that between 1056 − 35.28 and 1056 + 35.28 will be unwilling to pay more than $15.00 for the combo. We must convert these values into percentages of the total number of moviegoers.

   $$\frac{1056 - 35.28}{1200} = 0.8506$$

   $$\frac{1056 + 35.28}{1200} = 0.9094$$

   For a 95% confidence level, between 85 and 91 percent of moviegoers will be unwilling to pay more than $15.00 for the combo of a large drink and a large popcorn.

6. **above a 68 to pass, at least 95 for an A, and at least 108 for an A+**

Since the teacher sets the scoring curve along a normal distribution with 68% of the students

falling between a 56 and an 82, the mean score on the test is the midpoint of this range. The

middle of the range from 56 to 82 is $\dfrac{56 + 82}{2}$, or 69; therefore, in order to pass, the student

must get a score higher than a 69 on the test. If the mean score is 69, and one standard

deviation from this number was 56 or 82, then the standard deviation must be 82 − 69 =

13. Since a normal distribution curve results in 95% of the scores being within two standard

deviations from the mean, going up another 13 would give this point; 82 + 13 = 95, so to get

an A on the test the student needs at least a 95. Three standard deviations in this case would

bring the needed number up to 95 + 13 = 108, so in order to get an A+, the student would need

to get a 108 on the test.

7. **No, the selection is probably not truly random.**

First, find the probability that there is no repeat of an episode within a given week. By the

fundamental counting principle, the total number of possible sets of 10 episodes,

allowing for repeats, is $236 \cdot 236 \cdot 236 \cdot 236 \cdot 236 \cdot 236 \cdot 236 \cdot 236 \cdot 236 \cdot 236$, or 236^{10}.

The total number of possible sets of 10 episodes without any repeats is $236 \cdot 235 \cdot 234 \cdot$

$233 \cdot 232 \cdot 231 \cdot 230 \cdot 229 \cdot 228 \cdot 227$. (This can also be written as the permutation

$_{236}P_{10} = \dfrac{236!}{(236 - 10)!}$.) The probability that the 10 episodes shown in a given week includes no

repeats is $\dfrac{236 \cdot 235 \cdot 234 \cdot 233 \cdot 232 \cdot 231 \cdot 230 \cdot 229 \cdot 228 \cdot 227}{236^{10}}$, which equals

approximately 0.824.

Because the probability of no repeats within the 10 episodes is 82.4%, the probability of one or more repeats is 17.6%.

We want to determine the likeliness of zero repeats per experiment (week) in a sample set of 20 experiments (20 weeks of 10 episodes being randomly selected from 236 episodes). It may be easiest to view as comparing the sample proportion (experimental probability) of 0 to the theoretical probability of having a repeat within one of the 20 weeks.

The population proportion, p, is the same as the theoretical probability of a given week

including a repeated episode, 0.176. Because we are only interested in whether or not a given

week includes a repeated episode, this is a Bernoulli trial. The population size, 236^{10}, is far more

than 20 times the sample size of 20, so sample proportions should be normally distributed, with

a mean of 0.176 and a standard deviation of $\sqrt{\dfrac{p(1-p)}{n}} = \sqrt{\dfrac{0.176(0.824)}{20}}$, or about 0.085.

One standard deviation equals 0.085, so two standard deviations equal 0.170. A sample
proportion of 0 is more than two standard deviations (0.170) less than the mean (0.176).
The area under the normal distribution curve to the left of two standard deviations to the left
of the mean contains half of 5% of the area under the entire curve. So, the probability of a
sample proportion of 0 (0 weeks with repeats out of 20 weeks) is less than 2.5%. It seems very
unlikely that there would be no episode repeats within any of the 20 weeks by chance alone.
The television channel probably does not use a completely random selection method to choose
episodes to air each week.

8. **(a) 400, 1%; (b) No, the quality control tests would have to cost less than \$5,700 for it to be**
    worth it.
    (a) Out of 10,000 flashlights produced, 5%, or 500 flashlights, will be defective, and the other
         95%, or 9500, will be in perfect working condition. Of the 500 defective flashlights, 80%
         will be correctly identified by the quality control tests as defective.

         $0.80 \cdot 500 = 400$

         Of the 500 defective flashlights, 400 will be identified as defective and discarded. The
         other 100 defective flashlights will be sold.

         Of the 9500 properly working flashlights, 90% will be correctly identified as such.

         $0.90 \cdot 9500 = 8550$

         Of the 9500 properly working flashlights, 8550 will be sold, and the other 950 will be
         incorrectly identified as defective and discarded.

         Out of a batch of 10,000 flashlights, the factory will discard a total of 400 + 950 = 1350
         flashlights. They will sell the remaining 8650. Of those 8650 flashlights, 100 are actually
         defective, so the probability that a consumer purchasing a flashlight made in this factory
         gets a defective one is $100/8650 \approx 0.01$, or about 1%.

(b)   The factory loses $12 in potential profit per working flashlight they discard. They currently discard 950 working flashlights incorrectly identified as defective (a false positive) out of a batch of 10,000. If the improved quality control tests cut the number of false positives in half, then they would result in only 1/2 (950) = 475 working flashlights being discarded.

For a batch of 10,000 flashlights, the number of flashlights saved from incorrect disposal by the test improvement would be 950 − 475 = 475. The amount of money saved, in terms of potential profit, would be 475 · $12 = $5,700. However, for the improved tests, there would be a cost to the company of $8,000 per batch of 10,000 flashlights. The cost would be greater than the profits saved, so the quality control test upgrades would not be worthwhile—unless the reduction in the number of defective flashlights they sell also has a substantial positive impact on their profits.

REFLECT

Congratulations on completing Chapter 8!
Here's what we just covered.
Rate your confidence in your ability to

- Understand the difference between sample surveys, experiments, and observational studies, and identify methods that provide representative sample populations for each of these

    ① ② ③ ④ ⑤

- Find the standard deviation or estimated standard deviation, as appropriate, for a data set, and understand the relationship between standard deviations, mean, and the normal distribution curve

    ① ② ③ ④ ⑤

- Use proportional reasoning, given ratios in a large representative sample population, to make inferences about the target population

    ① ② ③ ④ ⑤

- Understand the difference between discrete uniform, continuous uniform, normal, skewed, and bimodal probability distributions

    ① ② ③ ④ ⑤

- Use area under a continuous probability distribution curve, either uniform or normal, to calculate probabilities, when appropriate to the situation

    ① ② ③ ④ ⑤

- Use simulations to assess how well experimental results match a given model and to determine when results are statistically significant

    ① ② ③ ④ ⑤

- Calculate the confidence interval and margin of error for a given sample proportion at various confidence levels and for various sample sizes

    ① ② ③ ④ ⑤

- Use statistics and probability concepts to interpret and evaluate given statements and to analyze options in real-world situations

    ① ② ③ ④ ⑤

If you rated any of these topics lower than you'd like, consider reviewing the corresponding lesson before moving on, especially if you found yourself unable to correctly answer one of the related end-of-chapter questions.

 Access your online student tools for a handy, printable list of Key Points for this chapter. These can be helpful for retaining what you've learned as you continue to explore these topics.

NOTES

NOTES

NOTES

NOTES

NOTES

NOTES

NOTES

International Offices Listin

China (Beijing)
1501 Building A,
Disanji Creative Zone,
No.66 West Section of North 4th Ring Road Beijing
Tel: +86-10-62684481/2/3
Email: tprkor01@chol.com
Website: www.tprbeijing.com

China (Shanghai)
1010 Kaixuan Road
Building B, 5/F
Changning District, Shanghai, China 200052
Sara Beattie, Owner: Email: sbeattie@sarabeattie.com
Tel: +86-21-5108-2798
Fax: +86-21-6386-1039
Website: www.princetonreviewshanghai.com

Hong Kong
5th Floor, Yardley Commercial Building
1-6 Connaught Road West, Sheung Wan, Hong Kong
(MTR Exit C)
Sara Beattie, Owner: Email: sbeattie@sarabeattie.com
Tel: +852-2507-9380
Fax: +852-2827-4630
Website: www.princetonreviewhk.com

India (Mumbai)
Score Plus Academy
Office No.15, Fifth Floor
Manek Mahal 90
Veer Nariman Road
Next to Hotel Ambassador
Churchgate, Mumbai 400020
Maharashtra, India
Ritu Kalwani: Email: director@score-plus.com
Tel: + 91 22 22846801 / 39 / 41
Website: www.score-plus.com

India (New Delhi)
South Extension
K-16, Upper Ground Floor
South Extension Part–1,
New Delhi-110049
Aradhana Mahna: aradhana@manyagroup.com
Monisha Banerjee: monisha@manyagroup.com
Ruchi Tomar: ruchi.tomar@manyagroup.com
Rishi Josan: Rishi.josan@manyagroup.com
Vishal Goswamy: vishal.goswamy@manyagroup.com
Tel: +91-11-64501603/ 4, +91-11-65028379
Website: www.manyagroup.com

Lebanon
463 Bliss Street
AlFarra Building - 2nd floor
Ras Beirut
Beirut, Lebanon
Hassan Coudsi: Email: hassan.coudsi@review.com
Tel: +961-1-367-688
Website: www.princetonreviewlebanon.com

Korea
945-25 Young Shin Building
25 Daechi-Dong, Kangnam-gu
Seoul, Korea 135-280
Yong-Hoon Lee: Email: TPRKor01@chollian.net
In-Woo Kim: Email: iwkim@tpr.co.kr
Tel: + 82-2-554-7762
Fax: +82-2-453-9466
Website: www.tpr.co.kr

Kuwait
ScorePlus Learning Center
Salmiyah Block 3, Street 2 Building 14
Post Box: 559, Zip 1306, Safat, Kuwait
Email: infokuwait@score-plus.com
Tel: +965-25-75-48-02 / 8
Fax: +965-25-75-46-02
Website: www.scorepluseducation.com

Malaysia
Sara Beattie MDC Sdn Bhd
Suites 18E & 18F
18th Floor
Gurney Tower, Persiaran Gurney
Penang, Malaysia
Email: tprkl.my@sarabeattie.com
Sara Beattie, Owner: Email: sbeattie@sarabeattie.com
Tel: +604-2104 333
Fax: +604-2104 330
Website: www.princetonreviewKL.com

Mexico
TPR México
Guanajuato No. 242 Piso 1 Interior 1
Col. Roma Norte
México D.F., C.P.06700
registro@princetonreviewmexico.com
Tel: +52-55-5255-4495
+52-55-5255-4440
+52-55-5255-4442
Website: www.princetonreviewmexico.com

Qatar
Score Plus
Office No: 1A, Al Kuwari (Damas)
Building near Merweb Hotel, Al Saad
Post Box: 2408, Doha, Qatar
Email: infoqatar@score-plus.com
Tel: +974 44 36 8580, +974 526 5032
Fax: +974 44 13 1995
Website: www.scorepluseducation.com

Taiwan
The Princeton Review Taiwan
2F, 169 Zhong Xiao East Road, Section 4
Taipei, Taiwan 10690
Lisa Bartle (Owner): lbartle@princetonreview.com.tw
Tel: +886-2-2751-1293
Fax: +886-2-2776-3201
Website: www.PrincetonReview.com.tw

Thailand
The Princeton Review Thailand
Sathorn Nakorn Tower, 28th floor
100 North Sathorn Road
Bangkok, Thailand 10500
Thavida Bijayendrayodhin (Chairman)
Email: thavida@princetonreviewthailand.com
Mitsara Bijayendrayodhin (Managing Director)
Email: mitsara@princetonreviewthailand.com
Tel: +662-636-6770
Fax: +662-636-6776
Website: www.princetonreviewthailand.com

Turkey
Yeni Sülün Sokak No. 28
Levent, Istanbul, 34330, Turkey
Nuri Ozgur: nuri@tprturkey.com
Rona Ozgur: rona@tprturkey.com
Iren Ozgur: iren@tprturkey.com
Tel: +90-212-324-4747
Fax: +90-212-324-3347
Website: www.tprturkey.com

UAE
Emirates Score Plus
Office No: 506, Fifth Floor
Sultan Business Center
Near Lamcy Plaza, 21 Oud Metha Road
Post Box: 44098, Dubai
United Arab Emirates
Hukumat Kalwani: skoreplus@gmail.com
Ritu Kalwani: director@score-plus.com
Email: info@score-plus.com
Tel: +971-4-334-0004
Fax: +971-4-334-0222
Website: www.princetonreviewuae.com

Our International Partners

The Princeton Review also runs courses with a variet
partners in Africa, Asia, Europe, and South America.

Georgia
LEAF American-Georgian Education Center
www.leaf.ge

Mongolia
English Academy of Mongolia
www.nyescm.org

Nigeria
The Know Place
www.knowplace.com.ng

Panama
Academia Interamericana de Panama
http://aip.edu.pa/

Switzerland
Institut Le Rosey
http://www.rosey.ch/

All other inquiries, please email us at
internationalsupport@review.com